AF758070

VIROLOGY RESEARCH PROGRESS

AVIAN INFLUENZA AND SARS

EPIDEMIOLOGY, GLOBAL PATTERNS AND CLINICAL MANAGEMENT

VIROLOGY RESEARCH PROGRESS

Additional books in this series can be found on Nova's website under the Series tab.

Additional e-books in this series can be found on Nova's website under the e-book tab.

VIROLOGY RESEARCH PROGRESS

AVIAN INFLUENZA AND SARS

EPIDEMIOLOGY, GLOBAL PATTERNS AND CLINICAL MANAGEMENT

HAILEY BENSON
EDITOR

New York

Copyright © 2015 by Nova Science Publishers, Inc.

All rights reserved. No part of this book may be reproduced, stored in a retrieval system or transmitted in any form or by any means: electronic, electrostatic, magnetic, tape, mechanical photocopying, recording or otherwise without the written permission of the Publisher.

For permission to use material from this book please contact us:
nova.main@novapublishers.com

NOTICE TO THE READER

The Publisher has taken reasonable care in the preparation of this book, but makes no expressed or implied warranty of any kind and assumes no responsibility for any errors or omissions. No liability is assumed for incidental or consequential damages in connection with or arising out of information contained in this book. The Publisher shall not be liable for any special, consequential, or exemplary damages resulting, in whole or in part, from the readers' use of, or reliance upon, this material. Any parts of this book based on government reports are so indicated and copyright is claimed for those parts to the extent applicable to compilations of such works.

Independent verification should be sought for any data, advice or recommendations contained in this book. In addition, no responsibility is assumed by the publisher for any injury and/or damage to persons or property arising from any methods, products, instructions, ideas or otherwise contained in this publication.

This publication is designed to provide accurate and authoritative information with regard to the subject matter covered herein. It is sold with the clear understanding that the Publisher is not engaged in rendering legal or any other professional services. If legal or any other expert assistance is required, the services of a competent person should be sought. FROM A DECLARATION OF PARTICIPANTS JOINTLY ADOPTED BY A COMMITTEE OF THE AMERICAN BAR ASSOCIATION AND A COMMITTEE OF PUBLISHERS.

Additional color graphics may be available in the e-book version of this book.

Library of Congress Cataloging-in-Publication Data

ISBN: 978-1-63463-793-0
Library of Congress Control Number: 2014958718

Published by Nova Science Publishers, Inc. † New York

Contents

Preface

Avian influenza (AI) is caused by Influenza A viruses, and are single stranded, segmented RNA viruses of the family *Orthomyxoviridae*. Influenza A viruses continue to pose a major threat to the poultry industry and to the public. Wild aquatic birds are considered the primary hosts of influenza A, in which the virus is enzootic. In these birds, influenza viruses usually replicate in the intestinal tract, cause no disease, and spread by fecal contamination of the water habitat. This book discusses the epidemiology, global patters, and clinical management of Avian Influenza. It also examines the symptoms, treatment and clinical outcomes of Severe Acute Respiratory Syndrome (SARS), which is a highly contagious respiratory disease caused by a SARS coronavirus.

Chapter 1 – An outbreak of any subtype of highly pathogenic (HP) or low pathogenic (LP) H5 or H7 avian influenza (AI) subtypes in poultry flocks from major poultry producing countries could have serious social and economic impacts. International trade will be disrupted and infected properties will need to be quarantined to control the disease. It is critically important that any suspicion of this disease is immediately reported to official local and national laboratories. If AI virus (V) is detected, it is important to eradicate the infection as quickly as possible to control further spread and reinstate the countries disease-free status for trade purposes. Comprehensive plans to respond to and control an outbreak of AI in poultry at a local and national level should be developed, if not already done, in consultation with major poultry industry groups. Agreed upon strategies should include: regular testing of free flying wild bird populations, commercial and backyard poultry flocks, and live bird markets for AIV infections; immediate stamping out and disposal of AIV infected birds to remove the major source of virus; strict quarantine and movement controls within a 9 kilometer radius to prevent the spread of infection; regular testing of flocks within this radius; decontamination to remove and reduce the virus in infected premises; tracing and surveillance to locate the source of infection; locate other infected premises and determine the extent of the infection; and zoning or compartmentalization to define infected and disease-free areas. Control of AIV without the use of vaccination is the preferred policy. Vaccination has not been necessary in most major poultry producing countries for AIV control, but its usefulness has been demonstrated in controlling Outbreaks in some countries, where Human and Veterinary infrastructure is lacking, farm biosecurity is lax, and there is large scale uncontrolled and untested backyard flocks and live bird markets (LBM), which contain multiple animal species, which lack sufficient clean up and testing. If the infection is spreading at a significant rate, ring vaccination around the infected area, enhanced biosecurity and other infection control measures should be implemented to protect flocks from infection.

Chapter 2 – Avian influenza virus (AIV) of subtype H9N2 cause serious economic losses and are of particular concern in Asia and Middle East due to severe outbreaks with high morbidity and mortality rates in poultry. As part of regular virological diagnostic sample investigations a current H9N2 virus A/chicken/Dubai/F5/2013 (DF5/13) was isolated from tracheal swabs. The virus was successfully propagated in embryonated chicken egg and MDCK cell culture system. Sequence analysis was carried out by amplifying the genes encoding the antigenic surface glycoprotein hemagglutinin (HA) and neuraminidase (NA) as well as the internal nonstructural protein (NS), matrix (M) and the nucleoprotein (NP). The gene sequence and phylogenetic analysis revealed that the current AIV isolate Dubai/H9N2 is genetically related to the G1 lineage of H9 viruses. It remained alone without forming any cluster with other closely related H9 viruses of the lineage G1. The nearest cluster or branch in the HA and NA phylogeny were the contemporary isolates from Israel and United Arab Emirates. Older Dubai AIV H9N2 isolates from chickens and quails dating from the year 2000-2003 deposited at the Gen Bank sequence database provided by the National Center for Biotechnology Information (NCBI) had maintained a clear distance cluster from the current AIV Dubai/H9N2 isolate and showed nucleotide divergence between 9.5 and 12.4%. The HA cleavage site of DF5/13 contained a ^{324}HARSSR*GLF^{332} motif correlating to the low pathogenic nature of the virus including an amino acid substitution at position 324 (P324H). The mentioned mutation and nucleotide divergence from older Dubai isolates points towards a genetic evolution of H9N2 viruses in Dubai. The recent DF5/13 isolate showed high infectivity titer on both mentioned propagation system. The replication kinetic described maximum titer within 32-48 hours post infection. The recent isolated H9N2 virus was found to be sensitive to amantadine, a M2 blocker drug which was further confirmed by *in situ* ELISA. This study emphasizes the need for continuous surveillance of H9N2 viruses in poultry in Dubai to obtain more information about the genetic evolution of these particular influenza viruses.

Chapter 3 – Chinese herbal medicine has been recorded to fight with infectious diseases for almost 2,000 years, and over 10,000 herbal medicines and 100,000 formulae are documented in ancient literatures. Chinese herbs not only stimulate, suppress or regulate the activity of immune system, but directly inhibit the proliferation or virulence of virus. In comparison with other anti-infection drugs, less side effect and mild cured process against infections are prominent advantages of Chinese herbal medicine. Since the outbreak of pandemic SARS, no pharmaceutical medicine even preventive vaccines have been deployed on human usage. Fortunately, a list of herbs has been proved to prevent and treat SARS coronavirus infections under different phases of trials, including human clinical trials. The extract of some Chinese herbs participates the coordination of human immune system, while some others or herbal compounds directly inhibit the activity of SARS-CoV. Thus herbal administrations are promising to manipulate this emerging infectious disease on basis of empirical, experimental and practical applications.

Chapter 4 – Severe acute respiratory syndrome coronavirus (SARS-CoV) is a novel virus that caused the first major pandemic of the twenty-first century. It was first reported in China in late 2002. The outbreak of severe acute respiratory syndrome in 2002–2003 exacted considerable human and economic burdens on countries involved. There has been no infection reported since April 2004. In affected countries, 8098 people have had probable SARS diagnosed, 774 of whom have died, with case-fatality rate of around 10%.

The diagnosis of SARS can`t be based on clinical features only but it needs both epidemiologic, and laboratory criteria that have been set by the World health organization and the Centers for Disease Control and Prevention. The clinical course of SARS appears to follow a triphasic pattern. The major clinical features on presentation include persistent fever, chills, myalgia, dry cough, sore throat, coryza, and extrapulmonary manifestation especially diarrhea. The radiographic appearances of SARS show predominant involvement of the lung periphery and the lower zone. The treatment of SARS can be with appropriate antibiotic coverage, antiviral agents and anti-inflammatory or immunomodulatory therapies. Mechanical ventilation can be used in cases with respiratory failure.

The prognostic factors associated with a poor outcome include age, chronic hepatitis B treated with lamivudine, high peak lactate dehydrogenase, high neutrophil count on presentation, presence of diabetes mellitus, or other comorbid conditions. A vaccine is not currently available in humans.

The prevention of SARS spread challenged government and health system due to physical stability of SARS-CoV in the environment, lack of protective immunity in the general population, and the absence of effective therapy or vaccination. Airborne precautions, inactivation of the virus using disinfectants, lung protective ventilation strategy, close observation to health care personnel caring for patient with suspected or confirmed SARS, and community spread controlling plans are our mean infection control weapons against SARS to prevent person-to-person transmission in future epidemics

In: Avian Influenza and SARS: Epidemiology ... ISBN: 978-1-63463-793-0
Editor: Hailey Benson © 2015 Nova Science Publishers, Inc.

Chapter 1

Strategies for Preventing and Controlling Avian Influenza Outbreaks in Poultry

Joseph J. Giambrone and Teresa V. Dormitorio
Poultry Science Department, Auburn University, Auburn, Al, US

Abstract

An outbreak of any subtype of highly pathogenic (HP) or low pathogenic (LP) H5 or H7 avian influenza (AI) subtypes in poultry flocks from major poultry producing countries could have serious social and economic impacts. International trade will be disrupted and infected properties will need to be quarantined to control the disease. It is critically important that any suspicion of this disease is immediately reported to official local and national laboratories. If AI virus (V) is detected, it is important to eradicate the infection as quickly as possible to control further spread and reinstate the countries disease-free status for trade purposes. Comprehensive plans to respond to and control an outbreak of AI in poultry at a local and national level should be developed, if not already done, in consultation with major poultry industry groups. Agreed upon strategies should include: regular testing of free flying wild bird populations, commercial and backyard poultry flocks, and live bird markets for AIV infections; immediate stamping out and disposal of AIV infected birds to remove the major source of virus; strict quarantine and movement controls within a 9 kilometer radius to prevent the spread of infection; regular testing of flocks within this radius; decontamination to remove and reduce the virus in infected premises; tracing and surveillance to locate the source of infection; locate other infected premises and determine the extent of the infection; and zoning or compartmentalization to define infected and disease-free areas. Control of AIV without the use of vaccination is the preferred policy. Vaccination has not been necessary in most major poultry producing countries for AIV control, but its usefulness has been demonstrated in controlling Outbreaks in some countries, where Human and Veterinary infrastructure is lacking, farm biosecurity is lax, and there is large scale uncontrolled and untested backyard flocks and live bird markets (LBM), which contain multiple animal species, which lack sufficient clean up and testing. If the infection is spreading at a significant rate, ring vaccination around the infected area, enhanced biosecurity and other infection control measures should be implemented to protect flocks from infection.

Introduction

Influenza A viruses continue to pose a major threat to the poultry industry and public health [1]. In fact, the Office International des Epizooties (OIE) has revised its international standard on AI concerned with the risk that a spread of AIV may present to the international community not only from economic, but also from public health points of view [2]. Uncontrolled AI can paralyze a nations industry and compromises its export of poultry. In addition, people in areas where AIV has shown to occur will often curtail the purchase of poultry products for fear of contracting an AIV induced disease. As part of the preparedness for AI outbreaks, it is important to maintain an active network that effectively monitors influenza activity in poultry across the country and the rest of the world, using the most versatile and fastest methods of detection available. This is so important because of the very rapid transit of people and farm animals across countries. Likewise, access to reliable vaccine strategies that can be quickly implemented to control the spread of AI during outbreaks is an important priority.

Intervention Strategies

AIV disease prevention and control involves 3 interrealted processes: bioexclusion, surveillence, and biocontainment [3]. Disease prevention is time consuming and expensive and requires total committment, because it normally involves eradication. Eradication is only effective if it has sufficient means of containing the infection followed by thorough cleanup preventing spread of the virus. These procedures require education of all personel involved in poultry production as well as sufficent infrastucture for the diagnosis of the disease and rapid quarantine, and "stamping out" of infected birds. Intensive educational programs are necessary to make owners, suppliers, wholesalers, dealers of commercial poultry and the general public aware of the threat of AI not only for poultry, but also potentially to humans. Education includes behavoir changes and communication. Farm workers need to be aware of how the virus in spread and these workers should not have access to backyard flocks or live animal markets. Education involves teaching and implamenting biosecurity, which involves modification in the way birds are reared, moved, sold, and cleaning and disinfection of the premises.

There is no single stratagey for AIV control and a combination of measures must be employed. In most countres infections are controlled within 3 months through quarentine and "stamping out" of infected flocks. However, in some developing countires, mostly in the Middle East, Asia, and Africa, this is not practical, so they limit infections through vaccination and depopulation and clean up of live animal markets. These markets are a continual source of infection for smaller less intensive farms as well as for exposure to humans. The virus is readily spread by free flying wild birds, which have easy access to these smaller farms. From there the virus can spread to large more intentsive poultry farms manily through the movement of commercial farm workers [1].

Epidemilogy

Wild aquatic birds (Order *Anseriformes* and *Charadriiformes*) are considered the primary hosts of influenza A, in which the virus is enzootic [4]. In these birds, influenza viruses usually replicate in the intestinal tract, cause no disease, and spread by fecal contamination of the water habitat [4]. These viruses occasionally infect land-based birds (Order *Galliformes)* causing epizootics. Numerous examples of transmission of influenza from ducks to other species have been recorded. However, these events remain relatively rare, considering the high frequency of exposure of land-based birds as a result of co-mingling with waterfowl shedding influenza. Other wild birds have also been shown to carry the virus. They include pigeons, sparrows, starling, quail, partridges, various parrot species, and finches [5]. However, these birds have little chance to enter the modern fully enclosed poultry houses, but holes in open sided curtan houses, which can occur, can allow small bird enterance.

In land-based poultry, influenza A infections range from asymptomatic to lethal. Virulence for chickens and other poultry can only result from infections with influenza A viruses that replicate efficiently in these species; i.e. that have crossed the species barrier. Viral determinants of virulence consist of individual viral components interacting with one another and with elements of the host cell, the immune system of the host, and the environment [6]. The importance of efficient replication and spread of influenza A virus in a new host, such as land-based poultry, cannot be overemphasized; it is a pre-requisite for the expression of virulence. Influenza viruses have been shown to infect all types of domestic or captive birds throughout the world. The probability of primary infections is proportional to the frequency of contact with feral birds. Provided that the virus can be transmitted horizontally in poultry, secondary virus spread is usually associated with human involvement, probably by transferring infective feces to susceptible birds.

Most efforts at controlling AI in poultry have been focused on the H5 and H7 subtypes; however, control measures should arguably be in place to diagnose and control AIVs regardless of the HA subtype. LP AI H9N2 and H6N1 strains circulating in LBMs served as gene donors in the reassortments leading to the emergence of the 1997 Hong Kong H5N1 in China [7]. This virus infected at least 18 people; killing six. The continuous circulation of LP non-H5, non-H7 LP AI in poultry markets of China has contributed to the subsequent outbreaks of highly pathogenic H5N1 viruses from 2001 to the present. Fortunately, these AIVs have not spread from person-to-person, although major surveillance efforts in LBMs in China continue to monitor closely the emergence of such a virus.

The highly lethal systemic disease is referred to as HP AIV. From the 1870s to 1981, HP has been known by a number of terms the most common of which is called fowl plague. Milder forms of LP AIV were first recognized between 1940s and the mid-1960s. Their impact on poultry production and trade has been much less severe than HP AIV.

Poultry markets around the world must be considered convenient sampling sites to detect the presence of AIVs, which can be devastating not only for poultry but also to humans. In the United States (US), LP AI H7 viruses circulating in the LBM system of the Northeastern States have undergone repeated reassortment with viruses of other subtypes [8]. It is possible that these reassortments may have contributed to successfully maintaining H7 viruses in the markets. These considerations underscore the importance of interspecies transmission of waterfowl AI infections to poultry, and alert to the large gap of knowledge regarding selection forces and molecular mechanisms that lead to the emergence of highly pathogenic viruses.

The LBM systems of the larger US cities and many other parts of the world are reputed to be mixing vessels for AIVs from commercial and non-commercial birds of various species. The largely uncharacterized links between the LBM system and commercial poultry have led to AI outbreaks in both. The LBM marketing systems have demonstrated that they embody the conditions under which dangerous viruses can emerge [9]. However, the specific practices that drive the emergence of viruses, which can cross species barriers, and can spread readily in commercial poultry are not well established. The existence of LBMs in many countries has been engrained in their culture for thousands of years and will undoubtedly exist for the foreseeable future. The populist of these ancient countries have an eversion for buying processed birds in large commercial markets.

Wild birds such as waterfowl and shore birds are major reservoirs for diverse populations of type A influenza viruses. Abundant epidemiological and virological data provide convincing evidence that influenza A viruses circulating in these wild bird reservoirs serve as parent strains for the HP and LP H5 and H7 influenza A viruses responsible for AI outbreaks in commercial poultry. Since the 1960s HPAI outbreaks have been documented in United Kingdom, Italy, Netherlands, Australia, Mexico, US, Canada, Hong Kong, China and Chile [1].

Transmission of AIVs to pigs and the ensuing roles that pigs may play in pandemic influenza virus creation (serving as hosts for adaptation of avian viruses to replication in mammals and for avian and human virus reassortment) have received extensive research attention [10]. However, virus transmission between birds and pigs is a bi-directional process. Pig-to-bird transmission poses economically important disease concerns for domestic turkeys and also raises the possibility for widespread distribution of influenza viruses across the human population via the migration of free-ranging waterfowl [11]. The turkey industry in the US includes production of over 300 million birds. Six LPAI outbreaks in turkeys in the US in 1978-2002 led to $179 million in losses [12]. Beyond such direct monetary costs, a potential mechanism for enhancing the infectivity of HP AI viruses for human beings is via initial replication in pigs [13]. Interestingly, turkeys could act also as amplifiers of influenza viruses that can transmit to humans as suggested [14].

Nonetheless, there has been very little research into the viral factors that control virus movement "backwards" from pigs to birds.

Surveillance

The establishment and maintenance of a coordinated, systematic type A influenza virus surveillance network in wild bird populations for countries must be done to prevent the entrance of these viruses into commercial poultry [1]. The surveillance network must consist of multi-participants using standardized protocols for sampling and analysis. This network should focus on non-migratory wild bird populations residing in areas along migratory waterfowl flyways. Non-migratory wild bird study sites should be selected as close as possible to the respective migratory bird study sites and, whenever possible, in close proximity to areas of significant LBM suppliers and/or commercial poultry production. Migratory birds should be sampled at the earliest opportunity upon first arrival at designated study sites, midway through local migration period and at the end of migration season for each respective study site. Non-migratory wild birds can be sampled as early as possible in the year and periodically during the year (as can be arranged), but at least one time before arrival of migrants, in the middle of migratory season, and after the migrants leave [4, 15].

Small back yard bird flocks, birds reared in free range, and LBMs should be regularly screened for AIV. In the US every commercial broiler, breeder, and table egg laying flock has to be tested at least once for the presence of AIV antibody and LBMs once per week. In addition, all day-old breeder chicks sent out of the US must also be examined for AIV antibody. If a flock is found to be infected with AI then all flocks within a predetermined distance should be regularly checked for AIV infection until the contaminated index flock has been cleaned and disinfected and repopulated.

All surveillance activities must include collection of ground-truth data using GPS data collection systems with database synchronization capabilities to Decision Support System. Antigenic and genetic characterization of AIVs isolated from any birds should be performed. If possible, pathogenesis studies should be performed *in vivo* on selected isolates in BSL3 animal holding facilities in order to characterize their ability to cross species barriers.

AIV Detection Methods

For the routine detection of AIV in clinically normal birds virus isolation in eggs and agar gel immunodiffusion assay are used. Although lengthy, they are considered the gold standards [16].

For surveillance and prevention and control of outbreaks; more rapid and highly sensitive methods are used. Nucleic acid-based detection (PCR) and protein-based detection (antibody or antigen) are available for early diagnosis of avian influenza.

Real time reverse transcription PCR (RRTPCR) has been used for the detection of influenza virus and it has been adopted by laboratories around the world as a key diagnostic test [17].

A great advantage of this system is that it can be modified to detect AIVs of all known subtypes and/or specific subtypes depending on the type of primers and probes needed. However, widely implementing this method in the LBM system and/or in small producer sites is difficult due to the cost of the equipment, reagents and the need for technically trained personnel. Recently developed multiplex (RRTPCR) have been developed to detect multiple HAs and NAs simultaneously [18].

There is an ever increasing number of commercially-available rapid influenza antibody-based detection kits [19]. These immunoassays are designed to detect influenza A virus and to provide a result within 10-30 minutes of set up; however they have been optimized to detect influenza (nucleoprotein or neuraminidase). The sensitivity of these tests are lower than for RRTPCR and they cannot discriminate among subtypes [20] (Table 1).

There is need to develop, validate, and implement in diagnostic laboratories across the world, more accurate, sensitive, reproducible and simpler tests specifically designed for early detection of AIVs (most notably H5, H6, H7, H9 and H10) subtypes, which have been associated with losses in poultry. Tissues for sampling from clinically normal birds include the trachea and cloaca, whereas sick birds may have virus in any organ showing lesions.

Table 1. Detection and isolation of avian influenza virus (AIV) and avian paramyxovirus (APMV) from cloacal swabs of hunter-harvested wild birds

Sample ID[1]	Sampling Date	Host[2]	HA titer	AC-ELISA	RRT-PCR-1[3] (Cp)	RRT-PCR-2[4] (Cp[5])	HI (NDV[6] titer)	Subtype(s)
GA26	1/7/06	NSHO	64	negative	Negative	36	negative	H10N7
AL121	1/6/07	RNDU	256	positive	Negative	24	negative	H1N1, H1N4
AL167	1/6/07	BWTE	256	positive	34.2	19	negative	H1N1
AL253	1/20/07	MALL	512	positive	Negative	20	negative	H1N1
AL299	1/27/07	AMWI	4	negative	Negative	40	negative	H1N1
AL381	1/21/07	AMWI	2048	negative	Negative	40	+ (640)	H1N1, APMV-1[7]
FL20	9/22/07	BWTE	256	negative	Negative	negative	+ (480)	ND[8]
FL21	9/22/07	BWTE	512	positive	27.4	12.4	negative	H3N8
FL28	9/22/07	BWTE	1024	positive	38.4	28.0	negative	H3N8
FL36	9/22/07	BWTE	1566	positive	33.6	22.4	negative	H3N8
FL43	9/22/07	BWTE	2048	positive	33.8	26.7	negative	H4N6
FL52	9/22/07	BWTE	384	negative	40.0	32.2	+ (1280)	APMV-1
FL55	9/22/07	BWTE	1024	positive	22.9	13.8	negative	H3N8
FL64	9/22/07	BWTE	1536	positive	Negative	16.8	negative	H4N6
FL77	9/22/07	BWTE	1536	positive	Negative	14.2	negative	H3N8
FL93	9/22/07	BWTE	128	positive	36.3	19.6	negative	H3N8
FL101	9/22/07	BWTE	128	positive	28.9	19.3	negative	H4N8
FL102	9/22/07	BWTE	1024	positive	34.2	18.7	negative	H3N8
FL105	9/22/07	BWTE	128	positive	36.6	15.6	negative	H3N8
FL107	9/22/07	BWTE	128	negative	40.0	15.9	negative	H4N8

Sample ID[1]	Sampling Date	Host[2]	HA titer	AC-ELISA	RRT-PCR-1[3] (Cp)	RRT-PCR-2[4] (Cp[5])	HI (NDV[6] titer)	Subtype(s)
FL108	9/22/07	BWTE	64	negative	38.4	32.3	+ (10)	APMV-4

Out of 825 virus isolation samples, 21 were HA + in allantoic fluid and further characterization of these 21 samples are shown. Hemagglutination (HA) titer, AC-ELISA, Real-time Reverse Transcription-polymerase Chain Reaction (RRT-PCR) of the AIV matrix gene, Hemagglutination Inhibition (HI) NDV titer, and subtyping test results. All tests except RRT-PCR-1 were conducted on allantoic fluids collected from swab inoculated embryonating eggs (passage 1). RRT-PCR-1 tests were done directly on swabs. AC-ELISA test results done directly on swabs of these 21 HA + samples were all negative and were not included in the table.

[1] ID = sample identification.

[2] NSHO = Northern shoveler, RNDU = Ring-necked duck, MALL = mallard, BWTE = blue-winged teal.

[3] RRT-PCR-1 = AIV detection directly from swab by real-time Reverse Transcription-Polymerase Chain Reaction.

[4] RRT-PCR-2 = AIV detection from allantoic fluids of swab inoculated embryonating egg by real-time Reverse Transcription Polymerase Chain Reaction.

[5] Cp = crossing point.

[6] NDV = Newcastle disease virus.

[7] APMV = avian paramyxovirus.

[8] ND = not determined.

Vaccination

Vaccination is the second line of defense against avian influenza. However, as it has been stated recently [21] there continues to be an argument about the use of vaccines for influenza control. Presently, inactivated, recombinant, and reverse genetic vaccines are in use around the world [22]. Other experimental vaccines, transgenic yeast vaccine and a DNA vaccine, have also been shown to be efficacious, but have not been commercialized [23, 24]. One important framework for this argument is whether it is better to use vaccines or to allow mildly pathogenic avian influenza to circulate unchecked. In the US only inactivated none H5 non-H7 AIV vaccines can be used and they require approval by the individual state governments; whereas H5 and H7 vaccines can only be used as part of an official USDA program and require approval by the federal government.

However, in some countries both inactivated recombinant vaccines using Newcastle diseases or Fowl Pox vector vaccines to control the morbidity and mortality due to HP H5 or H7 can be given by injection [21]. Although they are efficacious in decreasing clinical signs, they do not produce a sterilizing immunity and therefore AIV can be shed, albeit at a reduced rate, to other farms and the process of vaccination must be continued indefinitely. No country that continues to vaccinate has eradiated the virus from their poultry flocks. In other countries inactivated H9 or H10 vaccines are used to reduce LP infections. These infections often occur in birds experiencing immunosuppression or other respiratory diseases, which leads to more serious disease.

Genetic Resistance

Despite ongoing research to decrease resistance to infection, little progress bas seen reported [25]. Most work has attempted to develop birds that are resistant to infection using conventional breeding techniques. Only minor resistance has been shown selecting birds partially resistant to LP AIVs [26]. Birds selected to have high expression of a certain (Mx) gene were shown to have delayed mortality due to an HPH5-AIV, but significant mortality still occurred in these birds. In the many years of breeding birds resistant to any infectious disease has been done only with Marek's Disease virus. In contrast, selection for resistance birds to metabolic diseases such as tibial dyschondroplasia, ascites, and aortic rupture has been accomplished.

Education

Building effective education and biosecurity programs are essential for AIV prevention and control. The risk assessments are needed to address the specific requirements associated with the development of a comprehensive education, outreach, and biosecurity programs for LBMs, the producers of wild captive waterfowl, and for the producers of commercial poultry. Education and outreach efforts are needed to target different stages of understanding, address specific audience needs, as a targeted incremental approach that emphasizes genuine understanding gained over time. Educational materials for AI control including stand-up presentations, printed and on-line fact sheets, and CDs must be generated and delivered to targeted audiences including integrated and semi-integrated poultry meat and egg industry, allied industry representatives, independent poultry producers, poultry dealers and distributors, LBM owners and workers, game bird producers, poultry and waterfowl fanciers and 4-H club members.

Once these programs have been delivered in a state or other province, local veterinarians can receive the training required to conduct additional training sessions using the same template. The educational program delivered to this audience serves to establish local contacts with the poultry farmers and to establish a means through which real epidemiologic data can be collected. Quality assurance and regional biosecurity programs must be developed to specifically assist all poultry producers. These materials need to be integrated with those of other participating regions so that both general and region-specific concepts are covered. Post-intervention data on producers and retailers should be collected using written surveys and individual interviews. Data from as many sources as possible should be organized and analyzed using appropriate statistical methods. A web based source of information should be developed and maintained linkable from well-established, germane URLs. The site (s) should provide utilitarian, non-bias information, using an intuitive web interface. Textual content should be developed at the 8th grade reading level to support outreach for both lay and scientific communities.

Biosecurity

There are a number of important items, which must be included in any biosecurity plan. If possible all farms should be surrounded with a fence and locked gate. If not, then the entrance to the farm should contain a gate. All gates and locked door entrances to the house should contain a sign saying no entrance without expressed permission of the farm manager. If possible cars should not be allowed with 50 meters of the houses. Another prevention method would have a vehicle drive through dip with disinfectant or if not the tires should be sprayed with a disinfectant. All farm visitors should not have visited an AIV infected farm before coming to another non-infected farm. A 7 day waiting period should be taken if a visitor enters an AIV infected farm. In large multiage farms visitors are usually required to shower and dress in farm only clothes. On smaller farms, dressing in coveralls with plastic boots or rubber boats that can be dipped in disinfectant should be followed. Disinfectants in foot baths and vehicle dips should be regularly changed when the odor is gone and the fluid becomes cloudy. At the least visitors should ware disposable boots. All visitors should sign in, state their business, and indicate if they have visited other farms within the same day. Man is the most common biological vector of disease causing microorganisms in the poultry house.

Pets should never be allowed in the house and other farm animals should be kept at least 50 meters from the houses. Farm ponds for feeding livestock should be kept far from the houses to not allow wild free flying waterfowl to congregate near the farm. Back yard birds or individuals that rear them or frequent LBMs should not be allowed near the houses. Houses should be enclosed if possible; bird and rodent proof, and farms should be at least 2 kilometers from another.

Management procedures are also important in controlling or preventing AIV infections. Sanitize all equipment that circulates between houses. Discard used chick pads. Remove and record the number of dead birds several times per day. Methods for disposal will relate to location and environmental considerations. Standard practices include incinerating, composting fermentation, or freezing for use in rendering into feed ingredients. Broiler and pullet grow-out operations should be on an all-in, all-out basis. Move chicks in, move birds out, clean up and repeat. Allow 10-14 days between lots of broilers and more for breeders and layers. LBMs should be depopulated once per week and cleaned and disinfected. Follow approved practices with regard to disinfections, light, bird density, feeder and water spaces. When completed, every detail of management practices can be evaluated on a precise, scientific basis. Keep honest, accurate and usable records.

Be on the alert for signs of disease. Note abnormal feathering, abnormal behavior, and decreased activity, decreased feed intake, dust levels, ammonia, abnormal droppings, and caked litter. Have some knowledge of the "normal" mortality. Normal mortality for broilers in the US is 0.25% each for the first and second weeks, 2% for the next 4 weeks and 0.5% during the last week. Mortality for pullets should be 5% until housing in the laying facility, and then 1% per month. Necropsy a proportion of dead birds and look for a general condition, location, size and color of any lesions.

Mow grass and weeds around house and remove any debris, which could serve as a shelter for external parasites and rodents. Fill any holes around houses. Keep cats outside house to act as rodent control. If houses contain curtains that have holes they need be repaired as soon as possible to make sure wild birds and rodents are kept out of the house. Practice safe and effective external parasite control. Keep doors closed and locked at all times to prevent unwanted visitors. Add and use fans religiously, foggers and/or evaporative coolers, and adequate brooders to maintain fresh and comfortable air. Keep wateres clean and sanitized. Add biofilm cleaners under high pressure to the water lines to keep them clean and clear. Biofilm, produced by certain bacteria, is a microscopic framework of salts and polysaccharides that can attach to the inner surface of pipes. It can act as a shelter for bacteria and viruses. Biofilm can clog waterers, adhere to viruses, bacteria, mold, vaccines, drugs, and vitamins, and inactivate sanitizers. Regularly check all automatic equipment.

Disease Diagnoses

Despite all methods used to prevent AIV or other pathogens from entering the house, no house is disease proof. If the flock experiences any change in feed or water consumption, has uneven growth, lowered egg production, decreased egg shell quality, increased morbidity or mortality, especially showing respiratory and enteric systems signs, and lesions then the nearest company veterinarian and diagnostic lab should be notified.

AIV infections in susceptible birds can be highly variable with regard to their location, and severity, depending greatly on the host species and presence of secondary infections, and immunosuppression. HP AIV infections in poultry can result in morbidity and mortality reaching 100%. Birds may die quickly without showing clinical signs or show both respiratory and enteric signs and lesions. With HP infections birds may show multifocal necrosis and edema of the combs and wattles and subcutaneous hemorrhage of the feet. Birds may have central nervous system signs, somnolence, and drop in egg laying and shell quality, and not eat or drink. In LP AIV infections, birds may show only respiratory signs (labored breathing, sneezing, tracheal and mucus emanating from the eyes, nares, and mouth). There may be a drop in weight gain and feed conversion and lowed egg production and shell quality. Morbidity and mortality will be low. Gross lesions seen with the HP AIV include hemorrhages over the heart muscle, lungs, gizzard, pancreas, cecal tonsil etc. LP AIVs induce mucous in the trachea, lung, and air sacs [5, 6].

Disease produced by HP can be confused clinically with fowl cholera and vvNDV. Disease produced by LP AIVs can be confused with a number of respiratory diseases including mycoplasma, infectious bronchitis, Newcastle disease, and Laryngotracheitis. Laboratory techniques noted later in this article show how to definitively diagnose infections with AIV.

A company employee should take birds that appear normal, show clinical signs, or have recently died to the nearest diagnostic lab in an enclosed box or cooler along with water, feed, and litter samples. The samples and birds along with a history of the disease process should be provided in great detail to the lab admission personnel. If AIV is suspected by the farmer, company representative, or admission personal then great hast should be undertaken by the diagnostician to exam the samples as soon as possible in an isolated area as possible, with precaution taken to prevent the infection of the personnel with the virus.

Disease Reporting

Using a necropsy exam, the presence of gross and microscopic lesions, serology, viral isolation, and RTRPCR any positive AIV results should be immediately reported to the State Veterinarian. Their job will be to relay the results to the official authorized country health laboratory and have positive samples sent overnight to them for farther confirmative testing. If the AIV confirmation is made, along with the specific HA and NA and HP or LP results by the country's central lab then the State Veterinarian will contact all major poultry companies and human health officials if they deem that it is a threat to human safety.

All poultry companies will then initiate procedures as outlined in the previously published and accepted AIV manuals. These include but are not limited to farm quarantine and depopulation and depopulation of areas in close proximity to the farm. The quarantine should restrict movement of birds and other farms animal as well as human traffic if need be. It is up to the State Veterinarian's office and possibly the Medical's examiner offices to make sure that the procedures are followed and if human threat exists then Federal Government agencies as well.

A highly successful education and training program must be delivered in the major meat bird producing regions of the nation. This program teaches participants a highly effective method of in house euthanasia and composting. The most commonly used methods for these procedures will also be discussed later. There are several roadblocks that must be overcome in order to implement biosecurity programs that include both commercial and niche market poultry producers. One of those critical factors is the distrust that these groups feel for each other. However, once a common goal, i.e., AI control, is established, then it is much easier to work out the details of a plan. In the early development of a plan, it must be clear to all parties that they cannot economically live with AI. It is then important to gain agreements that there will be no punishment for people that conduct surveillance and find infections. Honesty among producers of all types is essential for successful AI control.

The success of any major program aimed at controlling a given infectious disease is measured by the prevalence of the agent at the beginning and at the end of the program. Working closely with the index and other AIV positive flocks, LBM system, and producers that supply those markets as well as the producers of wild captive birds, is needed to determine the risks factors that promote the perpetuation of AI in land-based poultry. The incidence of AIV in flocks and LBM systems and strategies to control the spread of and, perhaps, eliminate AI in these systems needs to be developed. Phylogenetic studies of any isolated viruses will provide evidence on the overall movement and spread of viruses in different states, countries, and across the globe.

Development and implementation of education programs (traditional- and electronic-based) will raise public awareness among market owners, suppliers, wholesalers, dealers of commercial poultry and the general public of the importance of AI for poultry and public health. Without scientific evidence it is difficult to change cultural and traditional management and market practices. This information will change the way the general public perceives AI and will help in the implementation of bioregulatory strategies that will limit the emergence of novel and more virulent influenza A virus strains.

Depopulation

Regardless of whether the flock needs to be immediately depopulated after AIV detection or allowed to go on for farther production, a thorough cleaning and disinfection procedure needs to be employed after the birds are gone. Depopulation of a non-AIV LBM system is less stringent than an AIV positive flock. Depopulation of AIV positive flocks includes rapid response management, culling, storage, transportation, composting, disposal, and repopulation. In an event of large magnitude of AIV outbreak, depopulated birds and other organic matter can be placed in land fill as approved by the State Veterinarian. Clear management structure includes a veterinary officer responsible for veterinary tasks, including, animal welfare, reporting, and clearance management. A non-veterinary field response manager is needed who is responsible for all non-veterinary tasks, including, cleaning, disinfection, and logistics. A health and safety office supervisor for the human health during the operation, including and occupation health and safety and gate management should also be present. An administration and finance officer supervising the financial management, registration, and communication processes should also be included. There should be clear, enforceable instruction instead of on-the-job training. In addition, each poultry company needs its own outbreak response plans.

Depopulation of an infected flock should be achieved within 24-48 hours. The most appropriate culling method should use humane killing according to a national approved method. It should utilize a limited number of operators and be easy to operate, thereby limiting exposure between responders, farmers, neighbors, pets, and nearby poultry farms. The technique that fits the outbreak situation plan for each farm location should be decided in advance.

In the past, the quickest and safest procedure for depopulation was to gather up the birds and eggs and place the mixture in a large container or truck and use CO_2 gas. However, this is laborious, time and labor consuming, and if the birds have HP H5 or H7 AIV, places the workers at risk. The latest techniques for mass euthanasia consist of anoxia using CO_2 foam from a fire truck. This is rapid and requires only a few individuals and does not require handling of the birds. Laboratory studies indicate that death occurs in a few minutes with little bird suffering [27]. This is now accepted by the USDA as the method of choice. However, work in Europe has shown that foam causes suffocation and bird suffering. They say that with this system the birds are killed by mechanical occlusion of the airway and it is unsuitable for caged birds. Their required method is to use high expansion nitrogen foam under a low pressure system. This foam kills by anoxia resulting in the total absence of O_2. The nitrogen and water/soap solution is mixed in a foam nozzle and delivered with a readily movable machine. It is suitable for all types of housing and animals. The authors are not familiar with this type of machine and wonders how quickly it can deliver foam to typically larger houses and farms, which exit in the US poultry industry compared to European farms.

Cleaning and Disinfection

Decontamination should than take place as quickly as possible to stop virus spread to other farms, prevent human infections, reintroduction to the same farms, and allow for safe repopulation. Regardless of how the birds are depopulated, the carcasses, dead birds, litter, eggs, feed, and other organic matter must be decontaminated before bringing the organic matter out of the house. A universally agreed upon method includes composting within the house for a minimum of 7 days. These systems will inactive AIV in the organic matter making it safer to remove from the house [28]. This composted decontaminated matter should than be taken by heaving machinery and buried in a nearby 2 meter deep pit as designated by the state veterinarian. During the process of depopulation and decontamination personnel should ware approved protective safety equipment, which includes an approved respirator, disposable coveralls, shoe covers, goggles, and disposable gloves. Disposable equipment should be incinerator if possible or buried with the composted material from the house, and other equipment should be treated with an approved disinfectant.

Removing all organic matter is necessary, because many disinfectants will not work under those conditions. There are a number of acceptable disinfectants that can be used but the most popular seems to be Verkon^R, (E. I. du Pont de Nemours and Company, Wilmington, Del.) which contains a sulfate compound. For mass application high pressure machines can be used. Scrubbing is essential and the chemical must be in contact with the surface for a minimum of 15 minutes to sterilize the surface.

Remove all unused feed and old litter and then clean feed and water equipment. Wash down houses with water and detergent. All dust, fecal material, and debris must be removed before disinfection will be effective. Disinfect and disinfest the house inside and out as much as possible. Disinfectants can be applied with a high pressure, hot water machine, with or without fumigation. For fumigation, use 35 cc of 37 % formalin, if allowed by

State and country governments, with 17.5 grams of potassium permanganate per 2.8 cubic meters of room space.

Water systems, lines, nipple diners, cups, or troughs should be flushed, sanitized and drained prior to removal from the house. Chemicals to remove biofilm and sanitize other drinking water contaminants include organic acids (Selko pH®, Selko, Inc. Tilburg The Netherlands) and hydrogen peroxide (CID 2000® CID LINESN. V. Waterpoortstraat 28900 Ieper - Belgium CID LINES N. V. Waterpoortstraat, The Netherlands CID LINES N. V. Waterpoortstraat 28900 Ieper - Belgium CID LINES N. V. Waterpoortstraat 28900 Ieper – Belgium and Aquaclean®, Aquaclean Systems, Gainesville, Ga.). They should be left in the water for 24 hours after clean out and then flushed clean with regular water.

Bell drinkers should be removed and disassembled for removal of organic debris to permit proper cleaning and disinfection. Reservoirs should be flushed, sanitized and drained during the cleaning procedure. Feed remaining in the pans, feed lines, chains, augers, or hoppers should be removed placed on the floor for removal and burial with the composted litter. These lines and equipment must be removed or raised prior to removal of litter and floor material.

Dirt floors cannot be disinfected even with composting, but once the house is totally cleaned and disinfected, the house can be heated to 40^0C for 24 to 48 hrs, along with a 30 day layout time before repopulation will help reduce organisms on the top layer of dirt as well as other cracks in crevices in the house. Other practices to reduce microorganisms in the litter, by reducing pH, include adding phosphoric acid 0.5 kilograms per 1 m^2 or superphosphate 1 kg per 1 m^2 or 25 kilograms of aluminum sulfate per 300 square meters of floor space. Some commercial compounds include Sodium bisulfate (Poultry Litter Treatment®), PLT, Jones-Hamilton Co. Walbridge, OH, Acidifed clay (Poultry Guard®, Vinayak Corporation Monmouth Junction, NJ 08852), and aluminum sulfate (AL Clear®, General Chemical, Parsippany, New Jersey). These compounds will also reduce ammonia levels in the house and at higher levels bind soluble phosphorus in the litter. The chemicals should be lightly moistened to dissolve them. A biological litter treatment, Micro Treat P®, Weed in Environments, Inc. Woodstock, Ontario, Canada has a blend of nonpathogenic microbes and enzymes. It can improve litter quality and reduce ammonia.

Microbial and urease inhibitors such as para-formaldehyde flacks and volatile fatty acids are antimicrobial. Some of these include De-Odorase® (Alltech, Lexington, KY., YS-50® (Yucca Plant extract and Ultra Litter®

Treatment, Clear View Enterprises, Tontitown, AR). Heating the house to 40°C for 24 hours or 50°C for 4 hours can also reduce pathogenic microbes. Inside and outside of poultry houses and hatcheries should be treated with disinfectants, insecticides, and rodenticides.

The ventilation system is also important, however, it is more difficult and time consuming to clean. Fans, casings, motors, belts, curtains, ventilation pads, and lovers should be individually cleaned free of manure, debris, dust and feather prior to disinfection. Equipment such as thermostats, scales, time clocks, electric panels switches and light bulbs etc., may need to be individually wiped, cleaned sanitized and protected from more sever effect of cleaning such as high pressure sprayers and disinfectant chemicals and protected from recontamination during the cleaning process.

Slates should be scraped of adhering caked manure and debris and cleaned and disinfected with a high pressure machine. Egg belts, egg flats, egg buggies, and packing machines should have egg material and shell debris removed prior to washing and disinfection. In laying and breeding houses, the storage areas, mechanical equipment, supply rooms should be cleaned of material, debris, and supplies for proper removal of organic materials and disinfectants. Any remaining eggs can be composted with the litter and dead birds.

Any dust and cobwebs should be blown down or vacuumed prior to disinfection. Curtains should be exposed to permit correct cleaning and disinfection. Houses should be washed down inside and outside with high pressure water and detergent. Roof areas and eaves with holes or nesting areas of wild birds should be addressed. Disinfect bulk feed tanks after removal of all feed, with special reference to caked feed. Air out the building for at least 7 days. Allow sunlight to disinfect equipment and inside of house.

Fumigation of the interior of the house can be done. In the US and many other countries, formalin has been prohibited due to human health concerns. Gulteraldehydes or hydrogen peroxide can be substituted. Disinfectants can be applied by low pressure spray. A perimeter of 2 meters around the exterior of the house should be free of uncut grass, materials, and obstruction. Roof areas and eaves with holes or nesting areas of wild birds should be addressed. Areas for rodent entrances or penetration should be sealed. Disinfectants can be applied by low pressure spray, granules, baits, etc. There are dozens, which can be found on the world wide web.

Repopulation

Before repopulation on a previously infected farm, specific plans should be in place to determine the source of the virus and prevent reinfection of the flock. The chain of infection needs to be broken or the infection may reoccur. There must be at least 1 month between flocks and no flocks in the near vicinity should still have active infections. The longer the time frame between flocks the less likely there is for microorganisms to remain viable on the farm. All new bedding should be placed on the farm and in some instances unvaccinated sentinel birds can be placed on the farm for 1 month and tested for AIV infection using serological techniques. The new flock and any other flocks within the 10 kilometer quarantined area should have 3 subsequent flocks tested for AIV, before any movement of birds from the farm takes place.

Conclusion

AIVs of any pathogenicity in commercial poultry, back yard flocks, or LBMs are a serious problem that requires quick action to prevent the spread. The ability of the virus to increase in pathogenicity, cross species, change antigenicity, and cause human health problems is extremely problematic. Its presence in wild free flying birds that travel many miles and cross continents makes eradication impossible. Its control requires a myriad of action to avoid catastrophic losses.

References

[1] Swayne, D. E., Suarez, D. L., Sims, D. L. *Influenza.* In Glisson, J. R. et al. eds. Diseases of Poultry. Ames, Iowa; Wiley-Blackwell, 2013, 181-219.

[2] *OIE. Terrestrial Animal Health Code. Avian Influenza.* http// www.oie.int/index.php? id = 169&L = 0&htmlfile = chapitre_ 1.10. 4htm. Accessed:22 October 2012.

[3] Collet, S. R. *Principles of Disease Prevention, Diagnosis, and Control.* In Glisson, J. R. et al. Diseases of Poultry. Ames, Iowa; Wiley-Blackwell, 2013, 3-61.

[4] Stalknecht, D. E., Shane, S. M. Host range of avian influenza. *Virus Research,* 1988, 12, 125-141.

[5] Alexander, D. J. *Orthromyxovirus Infections.* In McFerran, J. B. & McNulty, M. S. Virus Infections in Birds. Elsevier Science, London, UK. 1993,287-316.

[6] Alexander, D. J. Avian influenza. Recent developments. *Vet Bulletin,* 1982, 52, 341-352.

[7] WHO. 2011. *Evolution of Avian H5 Hemagglutin.* http//www.whoint/influenza/gisrs_laboratroy/201101_h5n1evoconceptualdiagram.pdf. Assessed 22.October, 2012.

[8] Murphy, B. R., Webster, R. G. *Orthomyxoviruses.* In Fields, B. N., Knipe, & P. M. Holey, eds. Fields Virology. Lippincott-Raven. Philadelphia, PA, 1996, 1396-1445.

[9] Davidson, S. R., Eckroade; R. J., Ziegler, D. F. A review of the 1996-1998 Nonpathogenic H7N2 avian influenza outbreak in Pennsylvania. *Avian Dis.* 2003. 47, 823-827.

[10] Kuiken, T., van den Bend, R. D., Jackwood, M. P., Swayne, D. L. Comparative pathology of select influenza A virus infections. *Vet. Pathol,* 2010, 47, 893-914.

[11] Halverson, D. A., Kelleher, C. J., Swayne, D. L. Epizootiology of avian influenza. Effect of season on incidence in ducks and domestic turkeys in Minnesota. *Appl. Env. Microbiology.* 1995, 49, 914-919.

[12] Halverson, D. A. (2009). Prevention and management of Avian Influenza; Experiences from the United States of America. *Rev. Sci. Tech.* OIE.28, 359-369.

[13] Swayne, D. E. *Understanding the ecology and epidemiology of avian influenza viruses; implications of zoonotic potential.* In Brown, C. C. & Bolin, C. A. eds. Emerging diseases of Animals. ASM Press. Washington, DC. 2000. 101-131.

[14] Capua, I., Alexander, D. L. Ecology, Epidemiology and human health implications of avian influenza viruses: Why do we need to share genetic data? *Zoonosis Public Health* 2008. 55, 2-15.

[15] Dormitorio, T. V., Giambrone, J. J., Guo, K., Hepp, G. R. Detection and characterization of avian influenza and other avian paramyxoviruses from wild waterfowl in parts of the southeastern United States. *Poult Sci,* 2009.88, 851-855.

[16] Swayne, D. E., Senne, D. A., Suarez, D. L. *Avian Influenza.* In Dufour-Zavala, D. E. et al. Isolation and identification of avian pathogens.

American Association of Avian pathologist. Jacksonville, F. L., 2008, 128-134.

[17] Spakeman, E., Senne, D., Myers, T. J., Bulga, L. L., Garver, M. L., Perdue; M. L., Lohman, K. L., Daum, L. T., Suarez, D. L. Development of a real time reverse transcriptase PCR assay for type A influenza viruses. *J. Clinical Microbiology,* 2002, 40, 3256-3260.

[18] Lu, H., Xie, L., Liu, L., Lin L. Studies on multiplex RT-PCR for detection of avian influenza virus type A group and specific H5 and H7 subtypes. *J. Vet. Sci. Med,* 2013, (1), 5, 1-5.

[19] Woolcock, P. R., Cardona, C. J. Commercial immunoassay kits for the detection of influenza virus type A: Evaluation of their use in poultry. *Avian Dis,* 2005, 49, 477-481.

[20] Dormitorio, T. V., Giambrone, J. J., Guo, K., Hepp, G. R. Evaluation of field and laboratory protocols to detect avian influenza viruses in wild aquatic birds. *Poult Sci.* 2009. 88, 1852-1831.

[21] Swayne, D. E. Application for new vaccine technologies for the control of transboundary diseases. *Dev. Boil. Basel,* 2004.19, 219-228.

[22] Swayne, D. E., Pavade, G., Hamilton, K., Vallet, B., Miyagishima, K. Assessment of national strategies for the control high pathogenic avian influenza and low pathogenic notifiable disease in poultry, with emphasis on vaccine and vaccination. *OIE Rev. Sci. Tech,* 2011. 30, 839-879.

[23] Hongzhuan, W. K., Williams, S, R., Singh, K, S., Gunn, N. K., Singh, T., Oh, S-C., Giambrone, J. J. Yeast derived avian influenza virus hemagglutinin protein induced immune response in SPF chickens. *J. of Animal Sci. and Vet. Advances,* 2011, 10 (8) 999-1002.

[24] Hongzhuan, W., Gunn, K. S., Singh, S. R., Singh, N. K., S.-C., Giambrone, J. J. A DNA vaccine against an H1N1 avian influenza induced humoral and cell-mediated in SPF chickens. *J. of Biotech Research,* 2011, 3:27-36.

[25] Swayne, D. E., Radin, M. J., Hoepf, T. M., Slemons, R. D. Acute renal failure as the cause of death in chickens following intravenous inoculation with avian influenza virus A/chickens/Alabama/7395/75 (H4N8). 1994. *Avian Dis,* 39, 51; 157.

[26] Hunt, H. D., Jadhan, Swayne, D. E. Major histocompatibility complex and background genes in chickens influenced susceptibility to high pathogenic avian influenza virus. *Avian Dis,* 2010, 54, 572-575.

[27] *Government approves fire fighting foam for exterminating birds.* http://www.upc-online.org/winter0607/foam.html 2007.Winter, Poultry Press.

[28] Bendfeldt, E. S., Peer, R. W., Flory, G. A. In house composting as a rapid response to avian influenza. *Biocycle,* 47, 38-40.

In: Avian Influenza and SARS: Epidemiology ... ISBN: 978-1-63463-793-0
Editor: Hailey Benson © 2015 Nova Science Publishers, Inc.

Chapter 2

Genetic Evolution of Avian Influenza Virus of Subtype H9N2 in Dubai

Rokshana Parvin, Kristin Heenemann and Thomas W. Vahlenkamp
Institute of Virology, Center for Infectious Diseases,
Faculty of Veterinary Medicine, University of Leipzig, Germany

Abstract

Avian influenza virus (AIV) of subtype H9N2 cause serious economic losses and are of particular concern in Asia and Middle East due to severe outbreaks with high morbidity and mortality rates in poultry. As part of regular virological diagnostic sample investigations a current H9N2 virus A/chicken/Dubai/F5/2013 (DF5/13) was isolated from tracheal swabs. The virus was successfully propagated in embryonated chicken egg and MDCK cell culture system. Sequence analysis was carried out by amplifying the genes encoding the antigenic surface glycoprotein hemagglutinin (HA) and neuraminidase (NA) as well as the internal nonstructural protein (NS), matrix (M) and the nucleoprotein (NP). The gene sequence and phylogenetic analysis revealed that the current AIV isolate Dubai/H9N2 is genetically related to the G1 lineage of H9 viruses. It remained alone without forming any cluster with other closely related H9 viruses of the lineage G1. The nearest cluster or branch in the HA and NA phylogeny were the contemporary isolates from Israel and United Arab Emirates. Older Dubai AIV H9N2 isolates from chickens and quails dating from the year 2000-2003 deposited at the Gen Bank sequence database provided by the National Center for Biotechnology Information (NCBI) had maintained a clear distance cluster from the current AIV Dubai/H9N2 isolate and showed nucleotide divergence between 9.5 and 12.4%. The HA cleavage site of DF5/13 contained a ^{324}HARSSR*GLF^{332} motif correlating to the low pathogenic nature of the virus including an amino acid substitution at position 324 (P324H). The mentioned mutation and nucleotide divergence from older Dubai isolates points towards a genetic evolution of H9N2 viruses in Dubai. The recent DF5/13 isolate showed high infectivity titer on both mentioned propagation system. The replication kinetic described maximum titer within 32-48 hours post infection. The recent isolated H9N2 virus was found to be sensitive to amantadine, a M2 blocker drug which was further confirmed by *in situ* ELISA. This study emphasizes the need for continuous surveillance of H9N2 viruses in poultry in Dubai to obtain more information about the genetic evolution of these particular influenza viruses.

Introduction

Avian influenza (AI) is caused by Influenza A viruses, are negative sense, single stranded, segmented RNA viruses of the family *Orthomyxoviridae*. At present, on the basis of antigenic properties of the glycoproteins hemagglutinin (HA) and neuraminidase (NA) influenza A viruses are classified into 18 HA subtypes (H1-H18) and 11 NA subtypes (N1-N11) (Tong et al., 2013). Based on the *in vivo* pathogenicity index, avian influenza viruses (AIVs) of subtype H5 and H7 are classified as either highly pathogenic avian influenza (HPAI) or low pathogenic avian influenza (LPAI).

Avian influenza in poultry with subtype H9N2 has become geographically widespread after first isolation from turkeys in Wisconsin in 1996 and has also been endemic in poultry populations across Asia and the Middle East. In the mid-1990s, outbreaks of H9N2 influenza viruses occurred in Germany, Italy, Ireland, South Africa, USA, Korea and China in multiple avian species. These viruses fall into a number of genetically defined lineages, with the majority of those circulating in Asia belonging to three lineages G1, Y280 andKr-p96323 represented by the prototype viruses A/quail/Hong Kong/G1/, A/duck/Hong Kong/Y280/97 and A/Chicken/Korea/38349-p96323/96, respectively, (Guan et al., 1999; Peiris et al., 2001; Xu et al., 2007).

H9N2 viruses are classified as low-pathogenicity avian influenza viruses causing decreased egg production and mild to moderate infections but sometimes they cause high mortality and a marked reduction in egg production in association with co-infected with other agents (Nili et al., 2002, 2003; Brown et al., 2006). Human infection with avian influenza viruses has underscored the continuing threat of a new pandemic virus with global health implications. Documented cases of human infection with H9N2 AIVs, first detected in 1999 in Hong Kong and China (Guo, 2002), indicate that these viruses can be directly transmitted from birds to humans.

It is well established that reassortment between isolates from different host species can generate viruses with pandemic potential. Fusaro et al., 2011 identified four distinct and co-circulating groups (A, B, C and D) of H9N2 viruses in Middle East, each of which has undergone widespread inter- and intra-subtype reassortments, leading to the generation of viruses with unknown biological properties. The detection of avian influenza H9N2 viruses in domestic pigs and humans clearly creates opportunities for such reassortment events (Butt et al., 2005; Xu et al., 2004). In addition, the extensive co-circulation of H9N2 viruses with other AIVs is likely to generate appropriate conditions for the development of reassortant viruses. Indeed, inter-subtype reassortments between co-circulating H9N2 virus and highly pathogenic H5N1 or H7N3 viruses have been detected in Bangladesh (Parvin et al., 2014), China (Guan et al., 1999; Xu et al., 2007) and Pakistan (Cameron et al., 2000; Iqbal et al., 2009). A/quail/Hong Kong/G1/97 is thought to be the donor of 6 internal genes to the poultry and human H5N1 viruses isolated in 1997 (Guan et al., 2000).The circulation of H9N2 viruses throughout Eurasia, along with their ability to infect mammals and humans and the potential for future reassortments, clearly raise concern about their pandemic potential.

The genetic and antigenic evolution of H9N2 viruses is well documented in China; on the other hand information on the genetic properties of those H9N2 viruses circulating in Central Asia and the Middle East is rather limited (Aamir et al., 2007; Homayounimehr et al., 2010; Iqbal et al., 2009; Li et al., 2005; Perk et al., 2006, Soltaniavar et al., 2010). However, it is important to explore the genetic characteristics, as well as the spatial and evolutionary dynamics, of the AIV H9N2 lineages that co-circulate in Central Asia and the Middle East. Since 1998, AIV H9N2 infections have been reported from Middle Eastern countries, causing widespread outbreaks in commercial chickens. In Iran, the first outbreak of AIV H9N2, strain Ck/IR/ ZMT-101/98, has been reported from layer flocks in 1998 (Vasfi and Bozorgmehri, 2002). HPAI H5N1, clade 2.2.1 was introduced into Egypt presumably by wild birds in February, 2006 (Saad et al., 2007; Aly et al., 2008). In 2010, AIV H9N2 was also reported from Egyptian broiler and broiler breeder farms for the first time (Monne et al., 2013). So far, few studied have reported the presence of AIV H9N2 from poultry in the United Arab Emirates (UAE) (Aamir et al., 2007; Wernery et al., 2013; Fusaro et al., 2011). All of those reported cases have been associated with to establish their phylogenetic relationship to the other Asian H9N2 lineages. Thus continuous surveillance, reporting of new outbreaks and the molecular characterization of the isolated viruses from the outbreaks in the Middle East is necessary. In this study one recent isolate A/chicken/Dubai/F5/2013 (DF5) of AIV H9N2 from a poultry flock in Dubai was reported which further contributed to establish the genetic lineage and also showed the genetic evolution of the current Dubai isolate from previous isolates in that particular region.

Recent Diagnosis of AIV H9N2 in Dubai

As part of regular virological diagnostic sample investigations we received three cloacal swab samples in PBS and twelve serum samples from the commercial chicken flocks in Dubai. Among the swab samples the presence of influenza and new castle disease virus was initially confirmed in two separate samples. The third swab sample remained negative for the both virus investigation tests. The influenza positive sample was further taken for virus isolation. The virus isolation was carried out by inoculation of swab sample into embryonated eggs (≈ 10 days) through the allantoic cavity route and also in Madin-Darby Canine Kidney (MDCK-II) cell. The eggs allantoic fluid (AF) and cell culture supernatants (CCS) were collected and subjected to hemagglutination (HA) test. The HA positive samples were further confirmed for the presence of AIV H9N2 by the quantitative real time RT-PCR targeted for the M gene and by standard RT-PCR using gene specific primer. The full length nucleotide sequences were carried out of the hemagglutinin (HA), neuraminidase (NA), matrix (M) and non-structural (NS) genes of the Dubai isolate. The phylogenetic relation and molecular genetic analysis of 4 genome segments were described and compared with the other representative AIV H9N2 strains available in Gen Bank. Nucleotide sequence identity and distance was calculated to check for the homology between the isolates. The AIV H9N2 strains in each genetic analysis were chosen on the basis of their sequence homology, geographic origin, year of the isolation in order to follow the genetic evolution, and the representative H9 strains from the particular lineage.

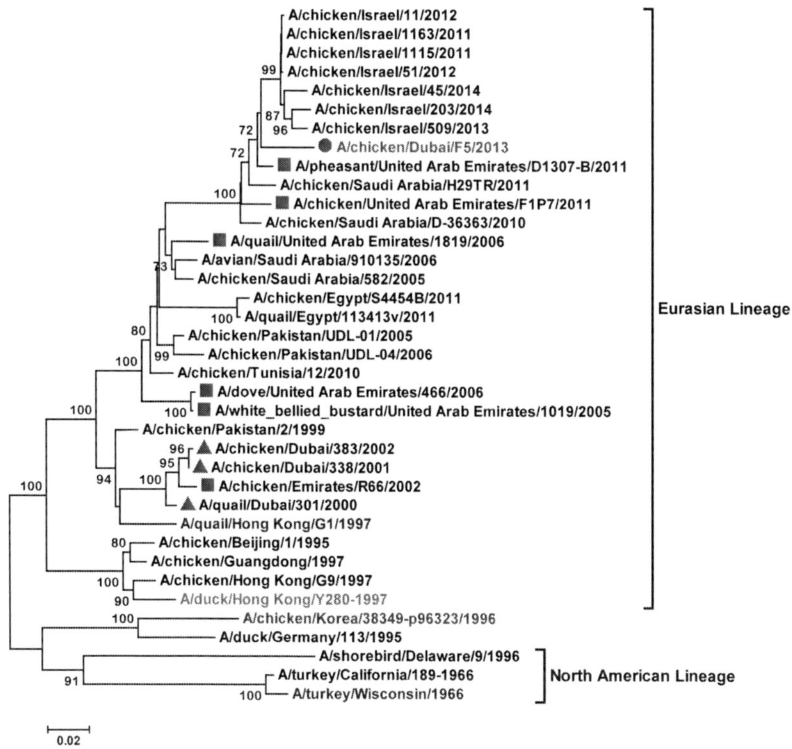

Figure 1. HA phylogeny; Phylogenetic relationship of the HA gene of AIV DF5-H9N2. The tree was generated by the distanced-based neighbor-joining method using MEGA 5.10. The reliability of the tree was assessed by bootstrap analysis with 1000 replications. Bootstrap supports > 70% are shown. Scale bar of each tree indicates substitution per site. Red round indicates the studied isolate. Red squires indicate AIV H9N2 strains from UAE. Red triangles indicate older Dubai isolates. The color taxon names indicate representative prototypes of the H9N2 lineages.

Phylogenetic and Homology Analysis of the Dubai H9N2

The sequence and phylogenetic analysis of HA and NA gene revealed that the current DF5-H9N2 is genetically related to the G1 lineage (Figure 1 and

Figure 2). Surprisingly, the currently isolated DF5 remain alone in the phylogenetic tree without forming any cluster with other Dubai strains and with other closely related AIV G1-H9N2 (Figure 1). The nearest cluster or branch in the HA and NA phylogeny were the contemporary isolates from Israel and UAE (Figure 1 and Figure 2). The H9N2 viruses that have reported from multiple avian species such as quail, dove and white bellied bustard in UAE at 2005-2006 may become the source of the current origin with a genetic evolution. The current AIV DF5-H9N2 maintained a clear distance with a significant nucleotide divergence between 9.5-12.4% (Table 1) compared to the older Dubai AIV H9N2 isolates in chickens and quails that originate from the year 2000-2003 (Aamir et al., 2007).

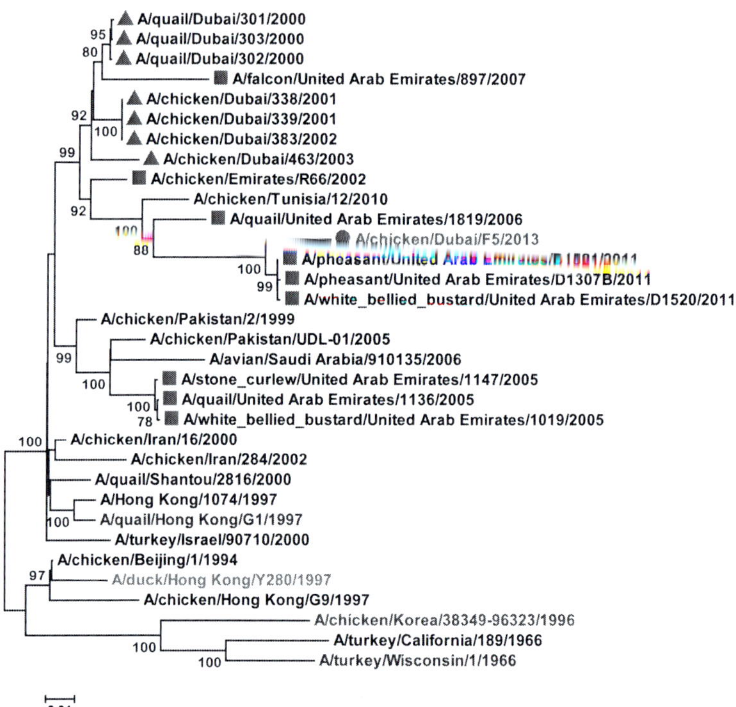

Figure 2. NA Phylogeny; Phylogenetic relationship of the NA gene of AIV DF5-H9N2. The tree was generated by the distanced-based neighbor-joining method using MEGA 5.10. The reliability of the tree was assessed by bootstrap analysis with 1000 replications. Bootstrap supports > 70% are shown. Scale bar of each tree indicates substitution per site. Red round indicates the studied isolate. Red squires indicate AIV H9N2 strains from UAE. Red triangles indicate older Dubai isolates. The color taxon names indicate representative prototypes of the H9N2 lineages.

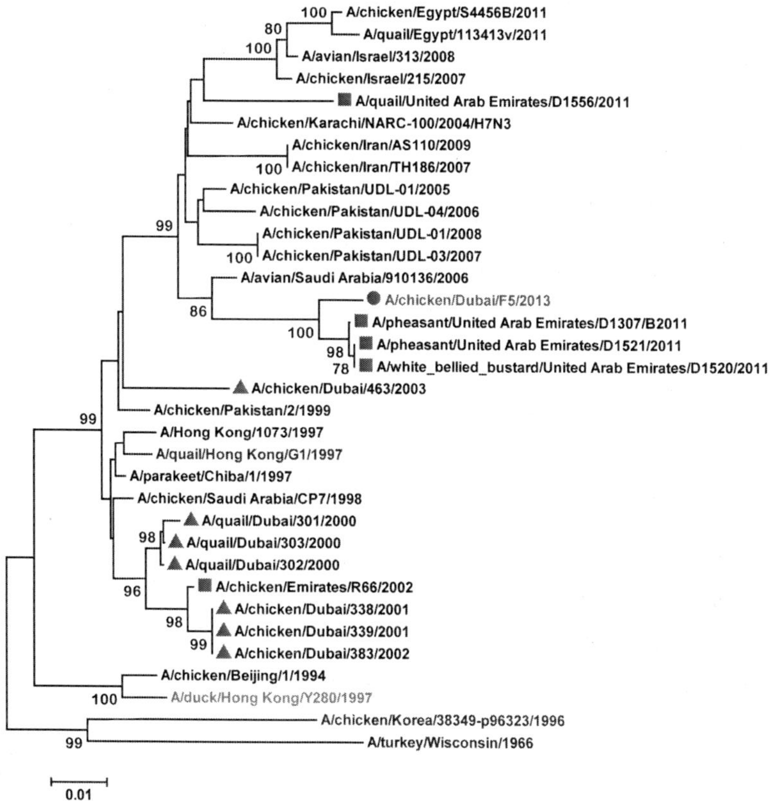

Figure 3. M phylogeny; Phylogenetic relationship of the M gene of AIV DF5-H9N2. The tree was generated by the distanced-based neighbor-joining method using MEGA 5.10. The reliability of the tree was assessed by bootstrap analysis with 1000 replications. Bootstrap supports > 70% are shown. Scale bar of each tree indicates substitution per site. Red round indicates the studied isolate. Red squires indicate AIV H9N2 strains from UAE. Red triangles indicate older Dubai isolates. The color taxon names indicate representative prototypes of the H9N2 lineages.

The internal M and NS gene phylogeny revealed, that the AIV DF5-H9N2 was closely related with the contemporary isolates of the avian species that were reported from UAE in 2011. However, unlike the HA and NA phylogeny the internal genes also remain alone without forming any cluster with other AIV H9N2 strains (Figure 3 and Figure 4). The nucleotide percent homology of the M and NS gene of DF5-H9N2 also varied compared to the other representative strains of H9N2 viruses (Table 1). The M gene maintained the prototype relation with G1 (Figure 3 and Table 1) whereas the NS gene followed the prototype A/turkey/Wisconsin/99-the first H9N2 isolate (Figure 4 and Table 1). Furthermore, the NS gene also showed genetic similarities with the other subtypes of avian influenza A viruses that are mostly isolated from wild bird species. Therefore, the recent Dubai H9N2 isolate exhibited a particular phylogenetic feature apart from other AIV H9N2 strains.

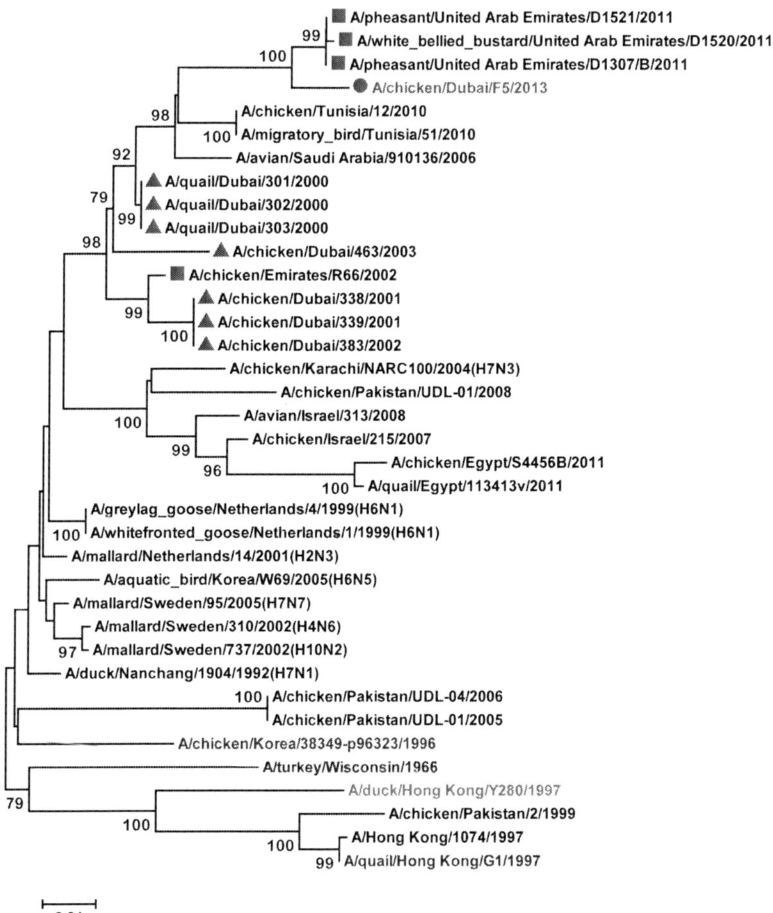

Figure 4. NS phylogeny; Phylogenetic relationship of the NS gene of AIV DF5-H9N2. The tree was generated by the distanced-based neighbor-joining method using MEGA 5.10. The reliability of the tree was assessed by bootstrap analysis with 1000 replications. Bootstrap supports > 70% are shown. Scale bar of each tree indicates substitution per site. Red round indicates the studied isolate. Red squires indicate AIV H9N2 strains from UAE. Red triangles indicate older Dubai isolates. The color taxon names indicate representative prototypes of the H9N2 lineages.

Table 1. Nucleotide present homology comparisons between the current Dubai isolate with the representatives prototypes and with other Middle-Eastern H9N2 isolates

H9N2 Strains	A/chicken/Dubai/F5/2013			
	HA	NA	M	NS
A/quail/Dubai/301/2000	89.1	90.3	93.7	95.5
A/quail/Dubai/302/2000	89.2	89.6	94.1	95.5
A/quail/Dubai/303/2000	89.1	90.2	94.1	95.5
A/chicken/Dubai/338/2001	88.9	89.8	93.5	93.5
A/chicken/Dubai/339/2001	88.9	88.4	93.5	93.5
A/chicken/Dubai/383/2002	88.7	89.8	93.5	93.5
A/chicken/Dubai/463/2003	88.9	90.0	94.1	93.6
A/pheasant/UAE/D1307B/2011	95.9	97.3	98.6	98.3
A/pheasant/UAE/D1521/2011	95.8	97.4	98.5	98.3
A/wbb/UAE/D1520/2011	95.7	97.3	98.5	98.2
A/chicken/Emirates/R66/2002	86.4	91.1	94.3	93.9
A/avian/SA/910136/2006	92.9	86.6	96.0	95.9
A/chicken/Egypt/S4456B/2011	88.9	89.5	93.6	89.5
A/chicken/Pakistan/2/99	88.1	89.0	95.2	86.8
A/chicken/Pakistan/UDL-01/2005	91.3	87.9	95.7	88.0
A/chicken/Pakistan/UDL-04/2006	92.1	88.7	95.3	88.0
A/quail/HK/G1/97	85.9	88.7	94.1	87.5
A/chicken/Korea/38349-p96323/96	58.1	82.0	88.8	89.9
A/duck/HK/Y280/97	85.9	86.3	92.1	88.0
A/turkey/Wisconsin/66	78.4	81.8	88.1	91.1

* Meg Align (pair distance Clusta lV analysis); wbb: white bellied bustard; UAE: United Arab Emirates; SA: Saudi Arabia; HK: Hong Kong.

Amino Acid Analysis of the Dubai H9N2

The HA, NA, M and NS genes play important roles in host-virus interactions and virulence (Knipe and Howley, 2007). Therefore, all the deduced amino acid sequences of the particular gene segments of the AIV DF5-H9N2 was determined and compared with other presentative strains of H9N2 viruses.

Table 2. Amino acid analysis and comparison of HA and NA gene of A/chicken/Dubai/F5/2013 H9N2 with representative H9N2 strains

Strains	HA				NA			
	cleavage site	RBS left edge	RBS right edge	Total glycosilation site	HB site 1st loop	HB site 2nd loop	HB site 3rd loop	Stalk deletion
	333-341	232-237	146-150		367-372	400-403	431-433	
A/ck/Dubai/F5/2013	HARSSRGLF	NGLIGR	GTSKS	29,105, 141, 298, 305 (5)	KRDLRA	SDDW	PRE	No
A/quail/Dubai/301/2000	PARSSRGLF	NGLQGR	GTSKA	29, 105, 141, 218, 298, 305 (6)	KKDLRA	SDNW	PQE	No
A/quail/Dubai/302/2000	PARSSRGLF	NGLQGR	GTSKA	29, 105, 141, 218, 298, 305 (6)	KKDLRA	SDNW	PQE	No
A/quail/Dubai/303/2000	PARSSRGLF	NGQQGR	GTSKA	29, 105, 141, 218, 298, 305 (6)	KKDLRA	SDNW	PQE	No
A/ck/Dubai/338/2001	PARSSRGLF	NGLMGR	GTSKA	29, 105, 141, 218, 298, 305 (6)	KKDLRA	SDNW	PQE	No
A/ck/Dubai/339/2001	PARSSRGLF	NGLMGR	GTSKA	29, 105, 141, 218, 298, 305 (6)	KKDLRA	SDNW	PQE	No
A/ck/Dubai/383/2002	PARSSRGLF	NGLMGR	GTSKA	29, 105, 141, 218, 298, 305 (6)	KKDLRA	SDNW	PQE	No
A/ck/Dubai/463/2003	PARSSRGLF	NGLLGR	GTSKA	29, 105, 141, 218, 298, 305 (6)	KKDLRA	SDNW	PQE	No
A/ph/UAE/D1307B/2011	HARSSRGLF	NGLIGR	GTSKS	29,105, 141, 298, 305 (5)	KRDLRA	SDNW	PRE	No
A/pht/UAE/D1521/2011	HARSSRGLF	NGLIGR	GTSKS	29,105, 141, 298, 305 (5)	KRDLRA	SDNW	PRE	No

Strains	HA				NA			
	cleavage site	RBS left edge	RBS right edge	Total glycosilation site	HB site 1st loop	HB site 2nd loop	HB site 3rd loop	Stalk deletion
	333-341	232-237	146-150		367-372	400-403	431-433	
A/wbb/UAE/D1520/2011	HARSSRGLF	NGLFGR	GTSKS	29,105, 141, 298, 305 (5)	KRDLRA	SDNW	PRE	No
A/ck/Emirates/R66/2002	PARSSRGLF	NGQLGR	GTSKA	29, 105, 141, 218, 298, 305 (6)	KEDLRA	SDNW	PQE	No
A/ck/SA/910135/2006	PARSSRGLF	NGLIGR	GTSKS	29,105, 141, 298, 305 (5)	KKDSRV	SDNW	PQE	No
A/ck/Pakistan/UDL-01/2005	PARSSRGLF	NGLIGR	GTSKS	29,105, 141, 298, 305 (5)	KKDSRA	SDNW	PQE	No
A/ck/Pakistan/UDL-04/2006	PARSSRGLF	NGLIGR	GTSKS	29, 105, 141, 298,305 (5)	KKDSRA	SDSR	PQE	No
A/ck/Egypt/S4456B/2011	PARSSRGLF	NGLIGR	GTSKS	29, 105, 141, 298,305 (5)	KKDSRA	SDSW	PQE	No
A/HK/1073/1999	PARSSRGLF	NGLQGR	GTSRA	29, 105, 141, 218, 298, 305 (6)	KKDSRS	SDNW	PQE	yes
A/ck/Korea/38349-p96323/96	PAASYRGLF	NGQQGR	GTSKA	29, 141, 218, 298, 305 (5)	SKDSRS	NNNW	PQE	No
A/quail/HK/G1/1997	PARSSRGLF	NDLQGR	GISRA	29, 105, 141, 218, 298, 305 (6)	KKDSRS	SDIR	PQE	yes
A/duck/HK/Y280/1997	PARSSRGLF	NGLQGR	GTSKA	29, 141, 218, 298, 305 (5)	KEDSRS	SDNW	PQE	yes
A/tk/Wisconsin/1966	PAVSSRGLF	NGQQGR	GTSRA	29, 141, 218, 298, 305 (5)	SKDSRS	SNNW	PQE	No

* Abbreviations; ck: chicken, ph: pheasant, tk: Turkey, HK: Hong Kong, UAE: United Arab Emirates, RBS: Receptor binding site and HB: Hemadsorbtion site.

* All amino acid numbering according to H9 numbering.

Table 3. Amino acid analysis and comparison of M and NS gene of A/chicken/Dubai/F5/2013 H9N2 with representative H9N2 strains

Strains	M							NS	
	M1	M2						NS1	
	15	26	27	30	31	34	38	Deletion (80-84)	C Domain
A/chicken/Dubai/F5/2013	I	L	V	A	N	G	L	No	GSEV
A/quail/Dubai/301/2000	I	L	A	A	S	G	L	No	GSEV
A/quail/Dubai/302/2000	I	L	A	A	S	G	L	No	GSEV
A/quail/Dubai/303/2000	I	L	A	A	S	G	L	No	GSEV
A/chicken/Dubai/338/2001	I	L	A	A	S	G	L	No	N/A
A/chicken/Dubai/339/2001	I	L	A	A	S	G	L	No	N/A
A/chicken/Dubai/383/2002	I	L	A	A	S	G	L	No	N/A
A/chicken/Dubai/463/2003	I	L	V	A	N	G	L	No	GSEI
A/pheasant/UAE/D1307B/2011	I	L	V	A	N	G	L	No	GSEV
A/pheasant/UAE/D1521/2011	I	L	V	A	N	G	L	No	GSEV
A/wbb/UAE/D1520/2011	I	L	V	A	N	G	L	No	GSEV
A/chicken/Emirates/R66/2002	I	L	V	A	S	G	L	No	N/A
A/chicken/Saudi Arabia/910135/2006	I	L	V	A	S	G	L	No	GSEV
A/chicken/Pakistan/UDL-01/2005	I	L	V	A	S	G	L	Yes	ESKV
A/chicken/Pakistan/UDL-04/2006	I	L	V	A	S	G	L	Yes	ESKV
A/chicken/Egypt/S4456B/2011	I	L	V	A	S	G	L	No	KSEV
A/Hong Kong/1073/1999	I	L	V	A	S	G	L	No	EPEV
A/chicken/Korea/38349-p96323/96	I	L	V	A	S	G	L	No	ESEV
A/quail/Hong Kong/G1/1997	I	L	V	A	S	G	L	No	EPEV
A/duck/Hong Kong/Y280/1997	I	L	V	A	S	G	L	No	EPEV
A/turkey/Wisconsin/1966	V	L	V	A	S	G	L	No	ESEV

*N/A: sequence not available; for Abbreviation see Tables 1 & 2.

The HA cleavage site, receptor binding site (RBS), amount of total glycosylation sites are important conserved region for the HA protein. The HA cleavage site is the primary virulence factor for influenza A virus which conserved region decide their pathogenicity (Knipe and Howley, 2007). The HA cleavage site of AIV DF5-H9N2 contained a ^{324}HARSSR*GLF^{332} motif confirming the low pathogenicity but also contained an amino acid substitution (P324H) at position 324 by the H9 numbering (Table 2). This specific mutation also observed in the AIV H9N2 strains from UAE in 2011 to which the current Dubai isolate maintained a close phylogeny relation as mentioned above. The receptor binding site and the total amount of glycosylation sites were also similar with the AIV UAE/2011 isolates. However, all the mentioned conserved regions did significantly vary with regard to the older DubaiH9N2 AIVs (Table 2). The NA stalk length correlates with the efficiency of virus replication and pathogenesis (Baigent and McCauley, 2003). Analysis of the stalk length revealed that the studied DF5-H9N2 did not show any deletions in contrast to G1 and Y280 and showed genetic similarities with older Dubai isolates in this particular conserved region. The binding pocket residues of NA hemadsorption site (HB) known for the active site for neuraminidase inhibitor drugs showed mutation in the 2nd and 3rd loop compared to all other H9N2 strains selected for the analysis (Table 2). Therefore, the response of the NA inhibitor drugs against DF5-H9N2 remains questionable.

The Isoleucine at position 15 of M1 protein showed conservation within all H9N2 lineages only differed from the first (A/turkey/Wisconsin/99) H9N2 strain (Table 3). Amantadine derivatives block the ion channel formed by M2 protein and inhibit the early step of virus replication .The substitutions of amino acid residue such as L26F, V27A, A30T, S31N, G34E and L38 of the M2 protein are already known to confer resistance to the amantadine (Belsche et al., 1988 and Hay et al., 1986). The current Dubai isolate exhibited mutation only at position 31 (S31N) like the isolates of UAE/2011. However, the DF-H9N2 was sensitive to amantadine which was further confirmed by the cellular ELISA in this study.

A previous study describes that the NS1 protein possess strain specific amino acid length (230-237) in poultry and mammalian host (Hale et al., 2008). The AIV DF-5 H9N2 exhibited full length NS1 (230) without any insertion and deletion. The C-terminal domain of the NS1 protein of avian influenza viruses contains PL motif as a potential virulence determinant. Large scale analysis of C-terminal domain from avian, human, swine and equine isolates have shown that each of them have atypical motif (Obenauer et al., 2006). The avian motif for many AIV H9N2 isolates varies due to frequent mutations (Parvin et al., 2014) which were also observed in this study (Table 3). However, the role of the PL motif in AIV H9N2 pathogenesis in poultry remains unknown. The above mentioned particular genetic constitutions and mutations of the DF-H9N2 virus confirmed that the unique H9N2 virus is circulating in the country and is significantly distinct from the older Dubai isolates. Therefore, genetic evolution occurred in the DF5-H9N2 virus, in contrast to the older Dubai AIV H9N2 strains. However, only one isolate does not represent the situation of all circulating H9N2 viruses of the country. Thus, further samples collection, isolation and characterization are recommended to confirm that, the older strains are still sustained or totally evolved.

Biological Properties of the Recent Dubai H9N2

The currently isolated and identified H9N2 virus from Dubai was propagated in embryonated chicken eggs (ECE) and MDCK-II cells. The allantoic fluids (AF) and the cell culture supernatants (CCS) were harvested at 2, 8, 16, 24, 32, 40, 48, 56 and 64 hpi and further subjected to virus titration. The growth kinetics of the AIV DF5-H9N2 based on different virus quantitation methods such as HA unit, $TCID_{50}$ assay and qRT-PCR target to M gene was investigated. The ECE-grown DF5-H9N2 virus showed high replication titer as the maximum titer reached up to about 8 log_{10} $TCID_{50}$/ml. Whereas, the cell culture-grown DF5-H9N2virus reached maximum yields up to 4 log_{10} $TCID_{50}$/ml which is four fold lower than the ECE-grown virus. On the HA test essay, the highest titers for the ECE-grown DF5-H9N2 was 9 log_2 (512) and for the cell culture-grown DF5-H9N2 was 7 log_2 (128), respectively. Thus, the ECE system allows better replication for the Dubai AIV H9N2 strain. The maximum genome copy number about 7 log_{10} genome copies/ reaction was obtained by the ECE-grown DF5-H9N2 virus earliest at 16 hpi. However, the maximum copy number obtained by the cell culture-grown DF5-H9N2 virus was about 6 log_{10} genome copies/reaction, detected at 48 hpi (Data not shown). Thus, the ECE system allowed rapid replication and yielded maximum titers within 16-32 hpi whereas, the cell culture system supported relatively slow replication. The avian influenza viruses preferentially bind to terminal α-2,3 SA linkage, whereas human influenza virus preferentially bind to α-2,6 SA. The allantoic cells of ECEs contain α-2,3 SA Gal while MDCK cells contain both linkages (Ito et al., 1997). Therefore, the specific receptor binding properties of embryos facilitate better replication of avian origin influenza virus (Rogers and Paulson, 1983). Epidemiologically, the DF5-H9N2 virus was known to induce clinical signs and caused mortality in commercial poultry flocks in Dubai. At the amantadine screening on MDCK-II cells at two different concentrations confirmed that the studied AIV DF5-H9N2 was sensitive to the amantadine M2 blocker drug, although the amino acid residue analysis of M2 protein showed a substitution (S31N). Therefore, single substitution may not be able to change the drug sensitivity of the studied H9N2 virus.

Conclusion

The current AIV H9N2 isolate from Dubai showed a particular feature both genetically and biologically. However, this isolate did not show any mutation that can alter virulence and thus remains non-pandemic for the human population. The rapid evolution of the H9N2 viruses leads to the possibility of new variants to emerge at any time. Therefore, the continuous monitoring and the extensive surveillance study from the particular regions are important and recommended.

References

Aamir U. B., Wernery U., Ilyushina N., and Webster R. G., 2007. Characterization of avian H9N2 influenza viruses from United Arab Emirates 2000 to 2003. *Virology* 361:45-55.

Aly M. M., Arafa A., Hassan M. K., 2008. Epidemiological findings of outbreaks of disease caused by highly pathogenic H5N1 avian influenza virus in poultry in Egypt during 2006. *Avian Diseases* 52: 269-277.

Baigent S. J., McCauley J. W., 2003. Influenza type A in humans, mammals, and birds: determinants of virus virulence, host-range and interspecies transmission. *Bio Essays* 25:657-671.

Belshe R. B., Smith M. H., Hall C. B. et al., 1988. Genetic basis of resistance to rimantadine emerging during treatment of influenza virus infection. *J Virol* 62:1508-1512.

Brown I. H., Banks J., Manvell R. J., Essen S. C., Shell W. et al., 2006. Recent epidemiology and ecology of influenza A viruses in avian species in Europe and the Middle East. *Developments in biologicals* 124: 45-50.

Butt K. M., Smith G. J., Chen H. et al., 2005. Human infection with an avian H9N2 influenza A virus in Hong Kong in 2003. *J Clin Microbiol* 43:5760-5767.

Cameron K. R., Gregory V., Banks J., Brown I. H., Alexander D. J., Hay A. J., Lin Y. P., 2000. H9N2 subtype influenza A viruses in poultry in Pakistan are closely related to the H9N2 viruses responsible for human infection in Hong Kong. *Virology* 278:36-41.

Fusaro A., Monne I., Salviato A. et al., 2011. Phylogeography and evolutionary history of reassortant H9N2 viruses with potential human health implications. *J Virol.* 85:8413-8421.

Guan Y., Shortridge K. F., Krauss S., Chin P. S., Dyrting K. C. et al., 2000. H9N2 influenza viruses possessing H5N1-like internal genomes continue to circulate in poultry in southeastern China. *J virol* 74:9372-9380.

Guan Y., Shortridge K. F., Krauss S., Webster R. G., 1999. Molecular characterization of H9N2 influenza viruses: were they the donors of the ''internal'' genes of H5N1 viruses in Hong Kong? *Proc Natl Acad Sci U.S.A.* 96:9363-9367.

Hale B. G., Randall R. E., Ortín J., Jackson D., 2008. The multifunctional NS1 protein of influenza A viruses. *J Gen Virol;* 89:2359-2379.

Hay A. J., Zambon M. C., Wolstenholme A. J. et al., 1986. Molecular basis of resistance of influenza A viruses to amantadine. *J Antimicrob Chemother* 18 Suppl B: 19-29.

Homayounimehr A. R., Dadras H., ShoushtariA., Pourbakhsh S. A., 2010. Sequence and phylogenetic analysis of the haemagglutinin genes of H9N2 avian influenza viruses isolated from commercial chickens in Iran. *Trop. Anim. Health Prod.* 6:1291-1297.

Iqbal M., Yaqub T., Reddy K., McCauley J. W., 2009. Novel genotypes of H9N2 influenza A viruses isolated from poultry in Pakistan containing NS genes similar to highly pathogenic H7N3 and H5N1 viruses. *PLoS One* 4(6):e5788.

Ito T., Suzuki Y., Takada A., Kawamoto A., Otsuki K., Masuda H. et al., 1997. Differences in sialic acid-galactose linkages in the chicken egg amnion and allantois influence human influenza virus receptor specificity and variant selection *J. Virol* 71: 3357-3362.

Knipe D. M. and Howley P. M., 2007. Orthomyxoviridae: The viruses and their replication. *Field's Vrology* 5th ed. v 2, 1647-1691.

Li C., Yu K., Tian G., Yu D. et al., 2005. Evolution of H9N2 influenza viruses from domestic poultry in mainland China. *Virology* 340:70-83.

Monne I., Hussein A. H., Fusaro A., Viviana V. et al., 2013. H9N2 influenza A virus circulates in H5N1 endemically infected poultry population in Egypt. *Influenza and other Respiratory Viruses* 7(3): 240-243.

Nili H. and AsasiK., 2003. Avian influenza (H9N2) outbreak in Iran. *Avian Dis.* 47 (3 Suppl.): 828-831.

Obenauer J. C., Dension J., Mehta P. K., Su X. et al., 2006. Large-scale sequence analysis of avian influenza isolates. *Science* 311:1576-1580.

Parvin R., Heenemann K., Halami M. Y. et al., 2014. Full genome analysis of avian influenza virus H9N2 from Bangladesh reveals internal gene reassortments with two distinct highly pathogenic avian influenza viruses. *Arch Virol* 159, 1651-1661.

Peiris J. S., Guan Y., Markwell D., Ghose P., Webster R. G., Shortridge K. F., 2001. Cocirculation of avian H9N2 and contemporary "human" H3N2 influenzaA viruses in pigs in southeastern China: Potential for genetic reassortment? *J Virol* 75:9679-9686.

Perk S., Banet-Noach C., Shihmanter E., Pokamunski S. et al., 2006. Genetic characterization of the H9N2 influenza viruses circulated in the poultry population in Israel. *Comp. Immunol. Microbiol. Infect. Dis.* 29:207-223.

Rogers G. N. and Paulson J. C., 1983. Receptor determinants of human and animal influenza virus isolates: differences in receptor specificity of the H3 hemagglutinin based on species of origin. *Virology* 127, 361-373.

Saad M. D., Ahmed L. S., Gamal-eldein M. A. et al., 2007. Possible avian influenza (H5N1) from migratory bird, Egypt. *Emerging Infectious Diseases* 13:1120-1121.

Soltaniavar M., Shoushtari H., Bozorgmehri fard M., CharkhkarS et al., 2010. Molecular characterization of hemagglutinin and neuraminidase genes of H9N2 avian influenza viruses isolated from commercial broiler chicken in Iran. *J. Biol. Sci.* 10(2):145-150.

Tong S., Zhu X., Li Y. et al., 2013. New world bats harbor diverse influenza a viruses. *PloS Pathog* 9: e1003657.

Vasfi M. M. and Bozorgmehri Fard M. H., 2002. Isolation of H9N2 subtypes of Avian Influenza viruses during an outbreak in chickens in Iran. *Iranian Biomed J* 6:13-17.

Xu C., Fan W., Wei R., Zhao H., 2004. Isolation and identification of swine influenza recombinant A/Swine/Shandong/1/2003(H9N2) virus. *Microbes Infect.* 6:919-925.

Xu K. M., Smith G. J., Bahl J., Duan L., Tai H. et al., 2007. The genesis and evolution of H9N2 influenza viruses in poultry from southern China, 2000 to 2005. *J Virol* 81(19):10389-401.

In: Avian Influenza and SARS: Epidemiology ... ISBN: 978-1-63463-793-0
Editor: Hailey Benson © 2015 Nova Science Publishers, Inc.

Chapter 3

Chinese Herbal Medicine: An Ancient Weapon on Emerging SARS

Xiaoguang Wang[1,], Shuai Bi[1], Fengxiao Zhu[2], Meiyi Xu[1] and Bei Zhao[1]*
[1]Institute of Microbiology, Chinese Academy Sciences, Beijing, P. R. China
[2]University College Dublin, Dublin, Ireland

Abstract

Chinese herbal medicine has been recorded to fight with infectious diseases for almost 2,000 years, and over 10,000 herbal medicines and 100,000 formulae are documented in ancient literatures. Chinese herbs not only stimulate, suppress or regulate the activity of immune system, but directly inhibit the proliferation or virulence of virus. In comparison with other anti-infection drugs, less side effect and mild cured process against infections are prominent advantages of Chinese herbal medicine. Since the outbreak of pandemic SARS, no pharmaceutical medicine even preventive vaccines have been deployed on human usage. Fortunately, a list of herbs has been proved to prevent and treat SARS coronavirus

* Corresponding author: Wang Xiaoguang, Institute of Microbiology, Chinese Academy Sciences, NO.1 Beichen West Road, Chaoyang District, Beijing 100101, P. R. China Email: wxg402@hotmail.com Telephone (FAX): +86 10 64806037.

infections under different phases of trials, including human clinical trials. The extract of some Chinese herbs participates the coordination of human immune system, while some others or herbal compounds directly inhibit the activity of SARS-CoV. Thus herbal administrations are promising to manipulate this emerging infectious disease on basis of empirical, experimental and practical applications.

Keywords: Traditional Chinese medicine, SARS, prevention, treatment

Introduction

With a long and well-recorded history, Chinese herbal medicine has been administrated to prevent and treat infectious diseases [1]. Nowadays Chinese herbal medicine is also popular in East and Southeast Asia as complementary therapies and has achieved remarkable success against infections. Over 10,000 herbal medicines and 100,000 formulae are documented in literatures, which supplies valuable resources to screen effective candidates on the purpose of infectious prevention and treatment [2]. In comparison with other anti-infection drugs, herbal therapies exhibit less side effect and mild cured process. Moreover, most Chinese herbs are safe for human administration according to long time practical applications, which would shorten the duration of new drugs' development, especially encountering new emerging infectious diseases.

Severe Acute Respiratory Syndrome (SARS), caused by a novel coronavirus, outbroke in China and spread to worldwide during 2002-2003. More than 8,000 SARS cases and 775 deaths were reported by WTO eventually. After the temporal disappearance in the summer of 2003, three new cases of laboratory-confirmed infection emerged in January 2004 [3]. The pathogenesis of SARS is associated with severe lung inflammations induced by excessive activation of immune system during the phase of viral infection [4]. Although it has been a decade since this pandemic infectious disease occurred, no pharmaceutical medicine or preventive vaccines has been employed on human usage. The development of SARS drugs is mainly focused on the target of immunomodulation or viral inhibitions. The recent emergence of Middle East Respiratory Syndrome (MERS) reminds us the severity caused by coronavirus, thus it is necessary to find effective ways to prepare for the return of SARS-CoV.

After the outbreak of SARS, Chinese herbs were used to prevent and treat SARS-CoV infections on basis of empirical remedies. During the peak of epidemic SARS, Chinese government suggested several herbal formulae on prevention, treatment and rehabilitation of SARS. According to the clinical reports, Chinese herbs and traditional medical physicians made a major contribution to combating SARS in China [5]. For instance, the incorporation of Chinese herbs alleviates clinical symptoms; facilitates the absorption of lung inflammation and pulmonary infiltration; reduces the risk of oxygen desaturation and abnormal fluctuation of oxygen saturation in the blood; reduces the dosage of glucocorticoid and antiviral agents [5]. Although Chinese herbs are effective on several clinical applications, the mechanism of these herbs on dealing with SARS-CoV is still inconclusive. Since clinical reports have been well reviewed elsewhere [3, 6-10], this chapter tries to explore how Chinese herbs or herbal compounds cope with newly emerging SARS.

Clinical Reports of SARS Cured by Chinese Herbal Medicine

Due to safe and efficacious record of administration, Chinese herbal medicine was promptly applied to cure SARS on basis of empirical and practical formulae. According to symptoms of SARS patients, different herbal formulae were distributed by herbal physicians. In the life-threaten circumstance, Western medicine was primarily incorporated in the treatment of SARS alone or with Chinese herbs. Clinical reports evidently showed that the combination of Chinese herbal medicine profoundly reduced SARS symptoms and increased health conditions of patients, reflecting the following aspects:

- decrease the dosage of glucocorticoid and antiviral agents
- alleviate clinical symptoms and improve quality of life
- facilitate absorption of lung inflammation and pulmonary infiltration
- reduce the risk of oxygen desaturation and stabilize oxygen saturation in the blood
- normalize levels of peripheral blood lymphocyte and increase levels of T cell subgroups

Furthermore, no mild or serious adverse effect was observed in the group integrated with Chinese herbs, indicating these herbal formulae were safe clinically. Representative clinical reports are documented and summarized in the following table.

Name	Formula I	Formula II	Formula III	Western medicine application	Chinese herbs plus Western medicine VS Western medicine alone	references
Report 1	*Gypsum fibrosum* (30-50 g), *Anemarrhena asphodeloides* Bge. (10 g), *Loincera japonica* Thunb. (30 g), Caulis Lonicerae Japonicae (30 g), *Forsythia suspense* (Thumb.) Vahl (10 g), *Notopterygium incisum* (10 g), *Mentha haplocalyx* Briq. (10 g), *Glycyrrhiza uralensis* Fisch. (6 g)	*Scutellaria baicalensis* Georgi (15-30g), *Gypsum fibrosum* (30 g), *Bambusa textilis* McClure (10 g), *Salvia miltiorrhiza* Bunge (15-30 g), *Panax pseudoginseng Wall.* var. notoginseng (Burkill) Hoo et Tseng (3 g)	*Pseudostellaria heterphylla* (Miq) Pax er Hoffm (15 g), *Scutellaria baicalensis* Georgi (15-30 g), *Atractylodes lancea* (Thunb.) DC. (10 g), *Atractylodes macrocephala* Koidz. (10 g), Dolicho LablabL. (30 g), Jobstears Seed (30 g), *Trichosanthes kirilowii* Maxim (10g), Vegetable Sponge of Luffa (10 g), *Salvia miltiorrhiza* Bunge (30 g), *Eriobotrya japonica* Thunb. (10 g)	Methylprednisolone (80-480 mg/d), ribavirin (1.0-2.0 g/d), azithromycin (0.5 g/d), levofloxacin (0.4 g/d), thymosin (160-200 mg/d)	Improve symptoms and pulmonary infiltration score, reduce the dosage and duration of corticosteroid, improve comprehensive index of SARS patients, no death case	[11]
Report 2	*Loincera japonica* Thunb. (20 g), *Isatis indigotica* Fortune(20 g), Rhizoma Cyrtomii Fortunei (15 g), Lobed Kudzuvine Root (15g), Folium Perillae (12g), *Platycodon grandiflorus* (Jacq.) A. DC. (10g), *Agastache rugosa*	Heartleaf Houttuynia Herb (45g), *Isatis tinctoria* (45 g), *Scutellaria baicalensis* Georgi (15 g), Amygdalus Communis Vas (15g), Radix Bupleuri (15 g), *Artemisia carvifolia* (15 g), *Agrimonia pilosa* Ldb. (20 g), *Caulis Bambusae* in Taeniam (15 g), *Gypsum fibrosum* (30 g), *Anemarrhena asphodeloides* Bge. (20 g), *Pseudostellaria heterphylla*	*Astragalus membranaceus* (Fisch.) Bunge. (45 g) *Glycyrrhiza uralensis* Fisch. (30 g), Semen Persicae (30 g), Carapax Amydae (30 g)	Methylprednisolone (80-320 mg/d), antibiotics, immunopotentiators	Reduce hospital stays and glucocorticoid dosage, inhibit recurrent fever	[12]

Name	Formula I	Formula II	Formula III	Western medicine application	Chinese herbs plus Western medicine VS Western medicine alone	references
	Fisch. et Mey.) O. Ktze. (15 g), *Glycyrrhiza uralensis* Fisch. (10 g)	(Miq) Pax er Hoffm (20 g), *Glycyrrhiza uralensis* Fisch. (30 g)				
Report 3	*Panax quinquenfolium* L. (15 g), *Ophiopogon japonicas* (Thumb Ker-Gawl.) (10 g), *Schisandra chinensis* (Turcz.) (15 g), Radix *Astragalus membranacers* (25 g)	Radix *Astragalus membranacers* (20 g), Radix *Codonopsis pilosula* (15 g), *Atractylodes macrocephala* Koidz. (10 g), *Poria cocos* (Schw.) (10 g), *Bupleurum chinense* DC. (15 g), *Paeonia lactiflora* Pall. (15 g), *Angelica sinensis* (30 g)	*Panax quinquenfolium* L. (20 g), *Adenophora stricta* Miq. (15 g), *Opiopogan japonicas* (Thunb.) Pyrostegia venusta (Ker-Gawl.) Miers (15 g), *Bupleurum chinense* DC. (20 g), *Scutellaria baicalensis* Georgi (10 g), Mulberry Leaf (15 g), *Lyrmeleon chinense* Mill. (15 g), *Attenmisa annua* L. (10 g)	Corticosteroid, antibiotics	Increase absorption of lung infiltration, reduce total dosage of corticosteroids, improve quality of life	[13]
Report 4		Chuan-Hu-Ning injection (0.8 g/d), Sheng-Mai injection (20-40 mL/d), Hu-Fei-Qing-Sha drink (150 mL), Jie-Du-Zhi-Tong capsule (3 tablets/d), Zhu-Yin-San-Jie capsule (3 tablets/d), Qing-Sha-Ling spray (16 sprays/d)		γ-globulin (10-20 g/d), *thymosin* (50-100 mg/d), meprednisone (40-100 mg/d), ribavirin (1 g/d),	Decrease the dosage of glucocorticoid and the incidence of complications	[14]

(Continued)

Name	Formula I	Formula II	Formula III	Western medicine application	Chinese herbs plus Western medicine VS Western medicine alone	references
Report 5	*Ephedra sinica* Stapf (5 g), Amygdalus Communis Vas (12 g), *Gypsum fibrosum* (45 g), *Anemarrhena asphodeloides* Bge. (10 g), *Lonicera japonica* Thunb. (15 g), *Forsythia suspensa* (Thunb.) Vahl (12 g), *Gardenia jasminoides* Ellis (12g), *Scutellaria baicalensis* Georgi (12 g), Folium Perillae (10 g), Artemisiacapillaris Thunb (15 g), Lobed Kudzuvine Root (15 g), *Pseudostellaria heterphylla* (Miq) Pax er Hoffm (15 g)	*Panax quinquefolius* L. (15 g), *Ophiopogon japonicas* (10g), *Schisandra chinensis* (10 g), *Cornus officinalis* Sieb. et Zucc. (12 g), *Aster tataricus L. f.* (15 g), *Eriobotrya japonica* Thunb. (12 g), Geosaurus (12 g), *Lepidium apetalum* Willd (15 g), *Salvia miltiorrhiza* Bunge (12 g), *Paeonia veitchii* (12 g), *Lonicera japonica* Thunb. (8 g), *Astragalus membranaceus* (Fisch.) Bunge. (10 g), *Trichosanthes kirilowii* Maxim. (15 g)	*Pseudostellaria heterphylla* (Miq) Pax er Hoffm (15 g), *Ophiopogon japonicas* (15 g), *Adenophora stricta* Miq. (15 g), *Atractylodes macrocephala* (15 g), *Eriobotrya japonica* (Thunb.) Lindl. (15 g), *Amomum villosum* Lour. (6 g), Jiao San Xian (30 g), *Astragalus membranaceus* (Fisch.) Bunge. (15 g), *Salvia miltiorrhiza* Bunge (15 g), Pericarpium Citri Reticulatae (6 g), *Polygonatum sibiricum* (15 g)	Corticosteroid (80-800 mg/d), ribavirin (800-1000 mg/d), IgA globulin (5-8 g/d), thymosin (100-160 mg/d), azithromycin (0.5 g/d), levofloxacin (0.4 g/d), moxifloxacin (0.4 g/d), ceftriaxone (2-4 g/d)	No death case, increase oxyhemoglobin saturation, reduce dosage of corticosteroid, no adverse effect	[15]
Report 6	*Gypsum fibrosum*, *Scutellaria baicalensis*, *Lithospermumery throrhizon* Sieb. Er Zucc	*Rehmannia glutinosa* Libosch, *Scrophularia ningpoensis* Hemsl.,. *Loinicera japonica* Thunb	*Pseudostellaria heterophylla* (Miq.) Pax er Pax er Hoffm, Radix Adenophora, *Salvia miltiorrhiza* Bunge	Methylprednisolone (80-160 mg/d), virazole，azithromycin, levofloxacin, Thymosins	Increase $CD4^+$ and total lymphocytes, improve symptoms and	[16]

Name	Formula I	Formula II	Formula III	Western medicine application	Chinese herbs plus Western medicine VS Western medicine alone	references
					absorption of lung filtrate	
Report 7	*Gypsum fibrosum* (45 g), Radix Bupleuri (15 g), *Anemarrhena asphodeloides* Bunge (10 g), Fritillary (10 g), *Scutellaria baicalensis* Georgi (15 g), *Artemisia carvifolia* (15 g), *Paeonia suffruticosa* Andr. (10 g), *Paeonia lactiflora* Pall. (12 g), *Forsythia suspensa* (Thunb.) Vahl (15g), *Cornus officinalis* Sieb. et Zucc. (30 g), *Atractylodes lancea* (Thunb.) DC. (15 g), *Agastache rugosa* （Fisch. et Mey.） O. Ktze. (10 g), Jobstears Seed (15 g), Amygdalus Communis Vas (10 g)	*Scutellaria baicalensis* Georgi (15 g), *Artemisia carvifolia* (15 g), *Trichosanthes kirilowii* Maxim. (30 g), *Salvia miltiorrhiza* Bunge (15 g), Flos Inulae Japonicae (10 g), *Curcuma petiolata* (10 g), *Acorus calamus* L. (10 g), Excrementum bombycis (15 g), *Atractylodes lancea* (Thunb.) DC. (15 g), *Atractylodes macrocephala* (15 g), *Polyporus umbellatus* (15 g), *Poria cocos* (Schw.) Wolf. (15g), Jobstears Seed (15 g), Amygdalus Communis Vas (10 g), *Plantaginis semen* (10 g), *Cornus officinalis* Sieb. et Zucc. (30 g),	*Panax quinquefolius* L. (30 g), *Astragalus membranaceus* (Fisch.) Bunge. (30 g), *Cornus officinalis* Sieb. et Zucc. (30 g), *Ophiopogon japonicas* (15 g), *Anemarrhena asphodeloides* Bunge (10 g), fritillary (10 g), *Thlaspi arvense* Linn. (30 g), *Forsythia suspensa* (Thunb.) Vahl (15 g), *Salvia miltiorrhiza* Bunge (15 g), Excrementum bombycis (15 g), Jobstears Seed (15 g), *Polyporus umbellatus* (15g), *Poria cocos* (Schw.) Wolf. (15g), *Trichosanthes kirilowii* Maxim. (30 g), *Aster tataricus* L. f. (15g)	Corticosteroid (80-640 mg/d), ganciclovir (500 mg/d), azithromycin (4 mg/d)+ levofloxacin (0.8 g/d), azithromycin (2 g/d) + Rocephin (8 g/d), Sulperazon (4 g/d) + azithromycin (2 g/d)	Improved the symptom score, reduce the number of days to loss of fever, improve absorption of pulmonary infiltration	[17]

(Continued)

Name	Formula I	Formula II	Formula III	Western medicine application	Chinese herbs plus Western medicine VS Western medicine alone	references
Report 8	*Gypsum fibrosum* (45 g), Radix Bupleuri (15 g), *Anemarrhena asphodeloides* Bunge (10 g), Fritillary (10 g), *Scutellaria baicalensis* Georgi (15 g), *Artemisia carvifolia* (15 g), *Paeonia suffruticosa* Andr. (10 g), *Paeonia lactiflora* Pall. (12 g), *Forsythia suspensa* (Thunb.) Vahl (15g), *Cornus officinalis* Sieb. et Zucc. (30 g), *Atractylodes lancea* (Thunb.) DC. (15 g), *Agastache rugosa* （Fisch. et Mey.） O. Ktze. (10 g), Jobstears Seed (15 g), Amygdalus Communis Vas (10 g)	*Scutellaria baicalensis* Georgi (15 g), *Artemisia carvifolia* (15 g), *Trichosanthes kirilowii* Maxim. (30 g), *Salvia miltiorrhiza* Bunge (15 g), Flos Inulae Japonicae (10 g), *Curcuma petiolata* (10 g), *Acorus calamus* L. (10 g), Excrementum bombycis (15 g), *Atractylodes lancea* (Thunb.) DC. (15 g), *Atractylodes macrocephala* (15 g), *Polyporus umbellatus* (15 g), *Poria cocos* (Schw.) Wolf. (15g), Jobstears Seed (15 g), Amygdalus Communis Vas (10 g), *Plantaginis semen* (10 g), *Cornus officinalis* Sieb. et Zucc. (30 g),	*Panax quinquefolius* L. (30 g), *Astragalus membranaceus* (Fisch.) Bunge. (30 g), *Cornus officinalis* Sieb. et Zucc. (30 g), *Ophiopogon japonicas* (15 g), *Anemarrhena asphodeloides* Bunge (10 g), fritillary (10 g), *Thlaspi arvense* Linn. (30 g), *Forsythia suspensa* (Thunb.) Vahl (15 g), *Salvia miltiorrhiza* Bunge (15 g), Excrementum bombycis (15 g), Jobstears Seed (15 g), *Polyporus umbellatus* (15g), *Poria cocos* (Schw.) Wolf. (15g), *Trichosanthes kirilowii* Maxim. (30 g), *Aster tataricus* L. f. (15g)	Thymosins (80 mg/d), Levofloxacin (0.4 g/d), Methylprednisolone （40-180 mg/d)	Increase $CD4^+$ and absorption of lung infiltrate, improve lung symptoms	[18]

Name	Formula I	Formula II	Formula III	Western medicine application	Chinese herbs plus Western medicine VS Western medicine alone	references
Report 9	Echinopslatifolius Tausch (15 g), *Forsythia suspensa* (Thunb.) Vahl (15 g), *Lonicera japonica* Thunb. (15 g), *Scutellaria baicalensis* Georgi (10 g), *Artemisia carvifolia* (15 g), *Gypsum fibrosum* (30 g), *Trichosanthes kirilowii* Maxim (15 g), Fritillaria thunbergii (10 g), *Citrus aurantium* L. (10 g), *Plantaginis semen* (10 g), *Paeonia veitchii* (12 g), *Gardenia jasminoides* Ellis (10 g)	Chinese Caterpillar Fungus, *Salvia miltiorrhiza* Bunge (15 g), Radix Pseudostellariae (20 g), *Scrophularia ningpoensis* Hemsl. (12g), *Paeonia veitchii* (12 g), EchinopslatifoliusTausch (15 g), *Forsythia suspensa* (Thunb.) Vahl (15 g), *Panax pseudoginseng Wall.* var. notoginseng (Burkill) Hoo et Tseng (3g), *Citrus aurantium* L. (10 g), Jobstears Seed (30 g)		Pefloxacin mesylate (0.4 g/d), azithromycin (1 g/day), Ribavirin (1.5 g/d), Thymosin (200 mg/d), Methylprednisolone (80-320 mg/d)	Improve absorption of pulmonary infiltration, increase clinical scores, no death case	[19]
Report 10	*Gypsum fibrosum* (16g), *Bupleurum chinenese* (12g), *Gardeniae jasminoides* (12 g), *Siler divaricatum* (10g), *Scutellaria baicalensis*	*Agaricus brazei* Murill, *Panax ginseng*, *Cordyceps sinensis*		Ribavirin (200 mg/d), Clarithromycin (250 mg/d), methylprednisolone (40 mg/d), immunoglobulin	Shorten the time of improvement	[20]

(Continued)

Name	Formula I	Formula II	Formula III	Western medicine application	Chinese herbs plus Western medicine VS Western medicine alone	references
	(10g), *Notopterygium incisium* (8 g), *Schizonepeta tenuifolia* (8 g), *Poria cocos* (8 g), *Paeoniae lactiflorae* (8 g), *Pinellia ternata* (8 g), *Agastache rugosa* (8 g), *Platycodon grandiflorum* (6 g), *Ophiopogon japonicus* (6 g)			(IVIG in short, Octagam, Gamimune, 2.5 g/50 mL)		

Table Brief clinical reports of Chinese herbs for SARS treatment (Only randomised controlled trials with herbal fomulae on SARS patients are included)

Alterations of Immune Systems

Chinese medical theories refer to the accordance of human body with nature and balance between inner system and pathogens. Strong evidence shows Chinese herbal medicine exhibits modulatory effects on human immune systems [21]. In consideration of the requirement of robust immunity against infections and fatal impact of excessive immune response caused by SARS-CoV, proper administration of effective herbs is beneficial to prevent and treat SARS during the epidemic. Despite that many Chinese herbs have immunomodulatory effect on human, this part of chapter only refers the effective herbs influencing immune responses under the SARS circumstance.

The extract of *Houttuynia cordata* Thunb. significantly stimulated the proliferation of mouse splenic lymphocytes in a dose-dependent manner [22]. The proportion of $CD4^+$ and $CD8^+$ T cells was also increased with the administration of the herbal extract. Moreover, IL- 2 and IL-10 secretions of mouse splenic lymphocytes were strongly elevated by herbal administration as well. In addition, the extract of *Houttuynia cordata* Thunb. exhibited significant dose-dependent inhibitory effects on SARS-CoV 3C-like protease ($3CL^{pro}$) and RNA-dependent RNA polymerase. Importantly, oral acute toxicity test demonstrated that the extract of *Houttuynia cordata* Thunb. was non-toxic to mice following oral administration at 16 g/kg. The lung fibrosis of SARS patients were partially relived following an oral administration of *Cordyceps sinensis* powder at 0.27g/day for 12 weeks [23]. On basis of murine bleomycin-induced lung fibrosis model, the administration of *Cordyceps sinensis* exhibited the reduction of infiltration of inflammatory cells, deposition of fibroblastic loci and collagen, formation of reactive oxygen species, and production of cytokines, as well as recovery from imbalance of MMP-9/TIMP-1 in preventive and therapeutic regimens [24]. In a fibrotic cell model with transforming growth factor beta-1 induction, the transition of human lung epithelial A549 acquired a mesenchymal phenotype and an increase of vimentin expression with a concomitant decrease of E-cadherin, were partially reverted by cordycepin, a major component of Cordyceps.

The herbal formula Kwan Du Bu Fei Dang [KDBFD] had immunomodulatory activities on murine splenic lymphocytes, showing elevation of $CD56^+$ NK cells and T-lymphocytes, including $CD8^+$ suppressor plus cytotoxic T-lymphocytes and $CD4^+$ helper T-lymphocytes [25]. Crude extract from *Coriolus versicolor* stimulated the proliferation of mouse splenic lymphocytes in a dose-dependent manner and increased the proportion of $CD4^+$ and $CD8^+$ T cells and levels of IL-2, IL-10, IL-4 and IFN-γ simultaneously. Extracts of *Ganoderma lucidum, Coriolus versicolor, Sinomenium acutum, Houttuynia cordata* and KDBFD inhibited SARS-CoV RNA dependent RNA polymerase in a dose-dependent manner and their IC_{50} values were 41.9, 108.4, 198.6, 251.1 and 471.3 μg/mL respectively. Moreover, the extract from *Houttuynia cordata* was confirmed to dose-dependently inhibit the activity of $3CL^{pro}$. An effective prophylaxis of SARS-CoV infections by Chinese herbs was deployed in Hong Kong during the peak of SARS. The administration of Yu-Ping-Feng and Sang-Ju-Yin significantly elevated the ratio of $CD4^{+}/CD8^{+}$ and reduced the number of B lymphocytes [26]. Interestingly, no SARS case was reported in the group administrated with Chinese medicine, however 0.14% (64/15,373) people without medicinal administration were diagnosed with SARS. The constituent of this formula included *Glycyrrhiza uralensis* Fisch., *Astragalus membranaceus* (Fisch.) Bge., *Forsythia suspensa* (Thunb.) Vahl, *Ledebouriella seseloides* Wolff, *Mentha haplocalyx* Briq., *Platycodon grandiflorus* (Jacq.) A. DC. and *Semen Armeniacae* Amarum. A multiherb remedy, containing Herba Houttuyniae, Flos Chrysanthemi Indici, Herba Artemisiae Scopariae, Herba Eupatorii, and Fructus Tsaoko, was recommended for the treatment and prevention of SARS by State Administration of TCM of the People's Republic of China during the SARS epidemic. Following analysis presented that the ethanolic extract of the five-herb remedy inhibited the classical pathway (CP) and alternative pathway (AP) of complement system *in vitro*, with CH_{50} and AP_{50} values of 0.83 and 1.24 mg/mL [27]. In use of bioactivity-guided fractionation and chromatographic separation, flavonoids exhibited significant inhibitory effect towards the CP and AP of complement system. Yin-Chiao-San (YCS), a traditional Chinese medicine, is widely prescribed to patients with pulmonary infectious diseases and was applied for the treatment of SARS in Asia. The YCS prescription mainly consists of ten Chinese herbs, including the Fructus of *Forsythia suspensa* (Oleaceae), the Flos of *Lonicerae Japonicae* (Caprifoliaceae), the Herba of *Lophatherum gracile* Brongn (Gramineae), the Radix of *Platycodon grandiflorum* (Campanulaceae), the Herba of *Menthae Haplocalycis* (Labiatae), the Fructus of *Arctium Lappa.* L (Compositae), the Radix of *Glycyrrhiza uralensis* Fisch (Leguminosae), the Semen of *Glycine* max (L.) merr. (Leguminosae), the Herba of *Schizonepeta tenuifolia* Briq (Labiatae) and the Rhizoma of *Phragmites communis* Trin (Gramineae). YCS exhibited antioxidant and anti-inflammatory activities and also inhibited collagen formation [28]. In comparison with bleomycin-treated mice, the incorporation of YCS significantly reduced the lung index, levels of malondialdehyde, hydroxyproline and tumor necrosis factor-α, and enhanced the catalase level in the lung. Furthermore the symptom of bleomycin-induced pulmonary fibrosis was alleviated in the presence of YCS.

Herbal Inhibitions on SARS-Cov

Some natural plants have antiviral activities to coronaviruses by directly inhibiting the infection or replication of virus [29-31]. It attracts us to screen effective herbs or natural compounds against infections caused by SARS-CoV. Due to effective complex herbs have been screened out by SARS patients clinically, the searching process for effective herbs or herbal compounds is shortened than ever before.

Complex Herbs against SARS

According to the synergistic theory of Chinese herbal medicine, mixed herbs are used to prevent or treat infection diseases. Three commercial Chinese herbal medicine, Fu-Fang-Lian-Pu granule (*Forsythia suspensa* (Thunb.) Vahl, *Taraxacum mongolicum* Hand.-Mazz., *Scutellaria baicalensis* Georgi, *Lonicera japonica* Thunb. and *Isatis indigotica* Fort.), Lian-Hua-Qing-Wen capsule (*Forsythia suspensa* (Thunb.) Vahl, *Houttuynia cordata* Thunb., *Lonicera japonica* Thunb., *Isatis indigotica* Fort., Ephedra sinica Stapf and *Glycyrrhiza uralensis* Fisch.) and Di-Kang injection (unknown components), effectively inhibited the activity of SARS-CoV, with EC_{50} values of 490, 110, 56 μg/mL respectively [32]. Ku-Gan electuary, containing extracts of *Ephedra sinica* Stapf., *Scutellaria baicalensis* Georgi, *Glycyrrhiza uralensis* Fisch., *Lonicera japonica* Thunb., *Mentha haplocalyx* Briq., *Platycodon grandiflorus* (Jacq.) A. DC. and *Semen Armeniacae* Amarum., could inhibit the replication of SARS-CoV based on Vero cell assay, with IC_{50} and TI values of 149 μg/mL and 9.12 respectively [33].

Consisting of extracts from *Lonicera japonica* Thunb. and *Bupleurum chinense* DC., An-Ti-Wei effectively inhibited the proliferation of SARS-CoV, with IC_{50} and TI values of 46.8 μg/mL and 8.0 respectively [34].

Single Herbs against SARS

Since hundreds of Chinese medicinal herbs are used to treat virus induced diseases, high throughput screening approaches are required to examine their antiviral activities against SARS-CoV. From a large library including more than 10,000 samples, effective products were screened on basis of cell assays [35]. Plant extracts of eucalyptus and *Lonicera japonica* were active against SARS-CoV at the concentration of 100 μM. By using MTS assay, a two-round screen showed *Lycoris radiata*, *Artemisia annua*, *Pyrrosia lingua*, and *Lindera aggregata* exhibited significant inhibition effects on virus-induced cytopathic effect [36]. The EC_{50} values of viral inhibition are 2.4±0.2, 34.5±2.6, 43.2±14.1, and 88.2±7.7 μg/mL respectively, and the CC_{50} values of these herbs were 690.5±21.0, 1022.9±55.1, 2127.3±178.9, and 1159.0±93.3 μg/mL respectively. Another screen of over 200 Chinese medicinal herbs was employed on anti-SARS-CoV activities by using a cell-based assay that measured SARS-CoV-induced cytopathogenic effect (CPE) *in vitro* on Vero E6 cells [37]. Six herbal extracts from *Gentiana scabra, Dioscorea batatas, Cassia tora, Taxillus chinensis* and *Cibotium barometz* were found to be potent inhibitors of SARS-CoV at concentrations between 25 and 200 μg/mL, with EC_{50} varied from 5 to 10 μg/mL. Furthermore, herbal extracts from *Cibotium barometz* and *Dioscorea batatas* significantly inhibited the activity of SARS-CoV $3CL^{pro}$ with IC_{50} values of 39 μg/mL and 44 μg/mL respectively. By screening 312 Chinese medicinal herbs, three widely used Chinese medicinal herbs of the family *Polygonaceae* clearly inhibited the interaction between SARS-CoV spike protein and host cell receptor angiotensin-converting enzyme 2 (ACE2) [38]. Furthermore, the IC_{50} values of herbal extracts from *Rheum officinale* Baill., *Polygonum multiflorum* Thunb. and *P. multiflorum* Thunb. ranged from 1 to 10 μg/mL. In comparison with other known antiviral Chinese herbs, herbal extract of *Rheum palmatum* L. significantly inhibited the activity of SARS-CoV $3CL^{pro}$ [39].

Bioactive Compounds against SARS

Bioactive compounds against SARS-CoV have been isolated from Chinese medicinal herbs recently. Baicalin derived from *Scutellaria baicalensis* and glycyrrhizin from *Glycyrrhiza uralensis* were effective against 10 clinical isolates of SARS-CoV on basis of neutralization test and plaque reduction assay [40]. Glycyrrhizin isolated from the root of *Glycyrrhiza glabra* was active in inhibiting replication of the SARS-associated virus (FFM-1 and FFM-2) [41]. Comparing to other commercial drugs, glycyrrhizin was the most effective drug against SARS virus and had the lowest toxic effect to human. The administration of glycyrrhizin also inhibited the adsorption and penetration of SARS-CoV, and was most effective compound during and after viral adsorption in Vero cell cultures. Moreover, some glycyrrhizin derivatives had much stronger anti-SARS-CoV activities but with increased cytotoxicity *in vitro* [42]. TSL-1, the extract from tender leaf of *Toona sinensis* Roem, exhibited an evident effect against SARS-CoV with selectivity index from 12 to 17 [43]. Another study showed quercetin, one compound of *Toona sinensis* Roem and structurally related to luteolin, had the inhibitory activity on the cellular entry of SARS-CoV, with the EC_{50} and CC_{50} of 83.4 μM and 3.32 mM respectively [44]. Ginsenoside-Rb1 isolated from *Panax ginseng* exhibited anti-SARS-CoV activity at the concentration of 100 μM [35]. Furthermore, both aescin from the horse chestnut tree and reserpine from genus *Rauwolfia* inhibited SARS-CoV replication at non-toxic concentrations. The EC_{50} values for Reserpine and Aescin were 3.4 μM and 6.0 μM respectively, and the corresponding CC_{50} values were 25 μM (SI 7.3) and 15 μM (SI 2.5). Lycorine isolated from *L. radiata* was proved to inhibit the CPE induced by SARS-CoV, with EC_{50} value of 15.7±1.2 μM [36]. Moreover the SI value of lycorine was more than 900.

Some bioactive compounds derived from Chinese herbs could directly inhibit the critical process of SARS-CoV. The extract from *Isatis indigotica* Fort. was used to prevent SARS-CoV infection during the outbreak of SARS in China due to empirical evidence. Subsequent study confirmed that the extract of *Isatis indigotica* Fort. directly inhibited the enzyme activity of SARS-CoV $3CL^{pro}$ [45]. Three components of *Isatis indigotica*, sinigrin, beta-sitosterol and indigo, dose-dependently inhibited cleavage activities of the $3CL^{pro}$ in cell-free and cell-based assays, with IC_{50} values of 217 μM, 1210 μM and 752 μM respectively. In the same research, two phenolic herbal derived compounds, aloe emodin and hesperetin, also inhibited the cleavage activity of the $3CL^{pro}$ *in vitro*, of which the IC_{50} were 366 μM and 8.3 μM respectively. The ethanol extract of *Torreya nucifera* leaves exhibited profound inhibition on SARS-CoV $3CL^{pro}$ activity (62% at 100 μg/mL) [46]. Further fractionation analysis showed eight diterpenoids and four biflavonoids effectively inhibited the activity of $3CL^{pro}$ in a dose-dependent manner. Within these bioactive compounds, the biflavone amentoflavone showed most potent $3CL^{pro}$ inhibitory effect with IC_{50} value of 8.3 μM. Two small molecules isolated from Chinese herbs, tetra-*O*-galloyl-β-D-glucose (TGG) and luteolin, were identified to exhibit prominent anti-SARS-CoV activity [44]. These natural molecules bound avidly with the surface spike protein of SARS-CoV and thus interfered with the entry of the virus to its host cells. TGG is a component from *Galla chinensis* and luteolin has been identified in extracts of many Chinese herbs such as *Veronica lina riifolia* Pall. Emodin, derived from genus *Rheum* and *Polygonum*, significantly blocked the interaction between SARS-CoV spike protein and ACE2 in a dose-dependent manner [38]. Furthermore the administration of emodin abolished the infectivity of spike protein-pseudotyped retrovirus to Vero E6 cells. Cinnamomi Cortex Extract (CCE) and Caryophylli Flos Extract (CFE) inhibited the activity of HIV/SARS-CoV S pseudovirus in a dose-dependent manner, with IC_{50} less than 100 μg/mL [47]. Furthermore the CCE and n-butanol extraction of CCE effectively inhibit the activity of SARS-CoV, however the CFE did not inhibit the infection of SARS-CoV. Procyanidin A2, procyanidin B1 and cinnamtannin B1 fractionated from CCE showed slight or moderate inhibitory effects in both HIV/SARS-CoV S pseudovirus infection and SARS-CoV plaque reduction assays. Moreover n-butanol extraction of CCE inhibited the internalization of TfR but the procyanidins and cinnamtannin did not, indicating unknown components of CCE interfered the endocytosis process. Artemisia annua is known as a herb medicine with therapeutic efficacies on antitracheitis, cough, phlegm and asthma. Bioinformatics analysis indicated Aurantiamide acetate derived from Artemisia annua shared a similar structure with MDL28170 which inhibited cathepsin L proteolysis and prevented SARS-CoV from entering a cell through membrane fusion [48].

Conclusion and Perspectives

Traditional Chinese medical theories mainly emphasize symptoms of infectious diseases, not causes [49]. The similar symptoms caused by unknown pathogens highlight the advantage of Chinese herbal medicine in the treatment of acute infectious diseases. During the outbreak of SARS, Chinese herbs played an indispensable role in prophylaxis and treatment of SARS-CoV infections via immunomodulatory pathway or direct inhibition on viral activities. Interestingly some Chinese herbs demonstrate dual functions on both immunomodulation and viral inhibitions. Since most herbal administrations have been proved to be efficacious and safe with long period practices, Chinese herbal medicine was promptly applied to clinic trials when encountering emerging SARS. The combination of Chinese herbal medicine and Western medicine exhibits advantages and efficacies over only Western medicine on the treatment of SARS [5], which is also confirmed by statistical evidence [50, 51]. More importantly, no or less side-effect has been observed in SARS patients received herbal therapies in a decade survey. However, the use of Western medicine against SARS causes short-term or long-term adverse effects [52, 53]. These prominent results of Chinese herbal medicine on treating SARS shed light on the remarkable efficacy and promising benefits of Chinese herbal medicine in coping with epidemics, pandemics or other unexpected infectious diseases.

However, the application and development of Chinese herbal medicine are confronting some bottleneck problems. The evaluation of Chinese herbal medicine by the model of conventional Western medicine is impractical and inappropriate because of different diagnostic and medicinal theories. For instance, Chinese herbal medicine emphasizes the whole herb or complex herbs on therapies, regarding as synergistic functions of different constituents. Therefore, it is difficult to identify specific compounds to particular infectious diseases. In the term of SARS, only few studies mentioned the mechanism of bioactive compounds against infections, and most results were published on Chinese journals in Chinese. It is necessary to publish results on international journals as well as increasing the quality of Chinese herbal research, which will share valuable experience of Chinese herbal therapies on SARS-CoV or SARS-CoV like viral infections. Additionally, the inadequate international standard for Chinese herbal medicine increases the difficulties of herbal applications around the world enduring SARS or SARS-like diseases.

References

[1] Wang, X. G. & Liu, Z. J. (2014). Prevention and treatment of viral respiratory infections by traditional Chinese herbs. *Chinese Med J, 127*(7), 1344-1350.

[2] Chen K. J. & Yu, B. (1999). Certain progress of clinical research on Chinese integrative medicine. *Chinese Med J, 112*(10), 934-937.

[3] Liu, J. P., Manheimer, E., Shi, Y. & Gluud, C. (2004). Chinese herbal medicine for severe acute respiratory syndrome: A systematic review and meta-analysis. *J Altern Complem Med, 10*(6),1041-1051.

[4] Chen, G., et al. (2004) Pathogenic role of HMGB1 in SARS? *Medical hypotheses, 63*(4), 691-695.

[5] World Health Organization. (2004). SARS : clinical trials on treatment using a combination of traditional Chinese medicine and Western medicine (World Health Organization, Geneva) p 196 p.

[6] Leung, P. C. (2007). The efficacy of Chinese medicine for SARS: a review of Chinese publications after the crisis. *Am J Chin Med*, *35*(04), 575-581.

[7] Zhang, M. M., Liu, X. M. & He, L. (2004). Effect of integrated traditional Chinese and Western medicine on SARS: A review of clinical evidence. *World J Gastroentero*, 10(23), 3500-3505.

[8] Chen, Y., Guo, J. J., Healy, D. P. & Zhan, S. (2007). Effect of integrated traditional Chinese medicine and western medicine on the treatment of severe acute respiratory syndrome: A meta-analysis. *Pharmacy Practice*, *5*(1), 1-9.

[9] Liu, X. M., Zhang, M. M., He, L. & Li, Y. P. (2012). Chinese herbs combined with Western medicine for severe acute respiratory syndrome (SARS). *Cochrane Db Syst Rev*, (10).

[10] Liu, X. M., Zhang, M. M., He, L., Li, Y. P. & Kang, Y. K. (2006). Chinese herbs combined with Western medicine for severe acute respiratory syndrome (SARS). *Cochrane Db Syst Rev*, (1).

[11] Hou, Y. J., et al. (2004). Integrated traditional Chinese and Western medicine for 34 patients with severe SARS. *Chin J Integr Tradit Western Med*, *24*(01), 81-82 (in Chinese).

[12] Li, J., et al. (2004). Clinical study on treatment of severe acute respiratory syndrome with integrative Chinese and Western medicine approach. *Chin J Integr Tradit Western Med* 24(1), 28-31, (in Chinese).

[13] Li, X., et al. (2004). Clinical study of integrated traditional Chinese and Western medicine for patients with severe acute respiratory syndrome on recovery stage. *Beijing J TCM*, 23(01), 26-27 (in Chinese).

[14] Li, Z. J., et al. (2003). Clinical study on treatment of severe acute syndrome with integrated traditional Chinese and Western medicin. *Chin J Integr Tradit Western Med Int Crit Care*, 10(04), 214-216 (in Chinese).

[15] Ren, A. M., Zhang, S. W., Wang, B. N., Liu, L. G. & Liang, X. M. (2004). Clinical study on treatment of severe acute respiratory syndrome by integrative Chinese and Western medicine. *Chin J Integr Tradit Western Med*, 24(2), 112-114 (in Chinese).

[16] Wang, R. B., et al. (2003). Preliminary study on clinical efficacy of integrative Chinese and Western medicine in treating severe acute

respiratory syndrome (SARS). *Chin J Integr Tradit Western Med*, 23(07), 492-493 (in Chinese).

[17] Zhang, X. M., et al. (2003). Clinical observation of effect of series of recipes 1, 2, 3 on SARS. *Chin J Traditl Chin Med Pharm*, 18(06), 323-325, (in Chinese).

[18] Zhao, C. H., et al. (2003). Randomized control study of integrated traditional Chinese and western medicine in treatment of 77 patients with severe acute respiratory syndrome. *Chin J Integr Tradit Western Med Int Crit Care*, 10(04), 197-200 (in Chinese).

[19] Zhang, L. S., et al. (2004). Integrated Chinese and Western medicine for 30 patients with SARS. *J Tradit Chin Med*, *45*(09), 675-677 (in Chinese).

[20] Hsu, C. H., et al. (2008). An evaluation of the additive effect of natural herbal medicine on SARS or SARS-like infectious diseases in 2003: A randomized, double-blind, and controlled pilot study. *Evid-Based Compl Alt*, 5(3), 355-362.

[21] Patwardhan, B. & Gautam, M. (2005). Botanical immunodrugs: scope and opportunities. *Drug discov today*, 10(7), 495-502.

[22] Lau, K. M., et al. (2008) Immunomodulatory and anti-SARS activities of *Houttuynia cordata*. *J Ethnopharmacol*, 118(1), 79-85.

[23] Hui, P. K. & Chan, M. H. (2006). *Cordyceps sinensis* improved post-SARS pulmonary fibrosis. *Hua Xia Med J (HongKong)*, 3, 172-176.

[24] Chen, M. L., et al. (2012). Protective roles of Cordyceps on lung fibrosis in cellular and rat models. *J Ethnopharmacol*, *143*(2), 448-454.

[25] Fung, K., et al. (2011). Immunomodulatory activities of the herbal formula Kwan Du Bu Fei Dang in healthy subjects: a randomised, double-blind, placebo-controlled study. *Hong Kong Med J*, 17, 41-43.

[26] Liu, D. H., Liang, B. Z. & Huang, L. Y. (2004). Clinical observation on the preventive effect of Kangdu Bufei decoction on acute severe respiratory syndrome. *Chin J Integr Tradit Western Med*, *24*(8), 685-688 (in Chinese).

[27] Zhang, T. & Chen, D. F. (2008). Anticomplementary principles of a Chinese multiherb remedy for the treatment and prevention of SARS. *J Ethnopharmacol*, *117*(2), 351-361.

[28] Yen, F. L., Wu, T. H., Liao, C. W. & Lin, C. C. (2007). A kampo medicine, Yin-Chiao-San, prevents bleomycin-induced pulmonary injury in rats. *Phytother Res*, *21*(3), 251-258.

[29] Jassim, S. A. A. & Naji, M. A. (2003) Novel antiviral agents: a medicinal plant perspective. *J Appl Microbiol*, *95*(3), 412-427.

[30] McCutcheon, A. R., et al. (1995). Antiviral screening of British Columbian medicinal plants. *J Ethnopharmacol*, *49*(2), 101-110.

[31] Vlietinck, A. J. & Vandenberghe, D. A. (1991). Can ethnopharmacology contribute to the development of antiviral drugs. *J Ethnopharmacol* *32*(1-3), 141-153.

[32] Zhu, S. Y., Li, X. Y., Wei, Y. L., Yang, P. Y. & qin, E. D. (2003). Inhibitory effects of three prescriptions of traditional Chinese medicine on SARS-associated coronavirus *in vitro*. *Lett Biotech*, *14*(05), 390-392(in Chinese).

[33] Wang, M. X., Sun, G. & Wang, X. H. (2004). Anti SARS virus by traditional Chinese medicine. *Chin J basic med tradit Chin med*, *10*(12), 38-39 (in Chinese).

[34] Cao, H. X., Wang, X. J., Yu, Y. H., Zhang, N. & Zhang, H. M. (2004). The pharmaceutical chemistry of Chinese medicine An-Ti-Wei in serum and its anti-SARS virus function. *Chin J Chin Mat Med*, 29(03), 93-94 (in Chinese).

[35] Wu, C. Y., et al. (2004). Small molecules targeting severe acute respiratory syndrome human coronavirus. *P Natl Acad Sci USA*, 101(27), 10012-10017.

[36] Li, S. Y., et al. (2005). Identification of natural compounds with antiviral activities against SARS-associated coronavirus. *Antiviral research*, *67*(1), 18-23.

[37] Wen, C. C., et al., (2011). Traditional Chinese medicine herbal extracts of *Cibotium barometz*, *Gentiana scabra*, *Dioscorea batatas*, *Cassia tora*, and *Taxillus chinensis* inhibit SARS-CoV replication. *J Tradit Complement Med*, *1*(1), 41-50.

[38] Ho, T. Y., Wu, S. L., Chen, J. C., Li, C. C., & Hsiang, C. Y. (2007). Emodin blocks the SARS coronavirus spike protein and angiotensin-converting enzyme 2 interaction. *Antiviral research*, *74*(2), 92-101.

[39] Luo, W. S., et al. (2009). Anti-SARS coronavirus 3C-like protease effects of *Rheum palmatum*, L. extracts. *Biosci Trends*, *3*(4), 124-126.

[40] Chen, F., et al. (2004). *In vitro* susceptibility of 10 clinical isolates of SARS coronavirus to selected antiviral compounds. *J Clin Virol*, *31*(1), 69-75.

[41] Cinatl, J., et al. (2003). Glycyrrhizin, an active component of liquorice roots, and replication of SARS-associated coronavirus. *Lancet*, *361*(9374), 2045-2046.

[42] Hoever, G., et al. (2005). Antiviral activity of glycyrrhizic acid derivatives against SARS - Coronavirus. *J Med Chem*, *48*(4), 1256-1259.

[43] Chen, C. J., et al., (2008). *Toona sinensis* Roem tender leaf extract inhibits SARS coronavirus replication. *J Ethnopharmacol*, *120*(1), 108-111.

[44] Yi, L., et al., (2004). Small molecules blocking the entry of severe acute respiratory syndrome coronavirus into host cells. *J Virol*, *78*(20), 11334-11339.

[45] Lin, C. W., et al. (2005). Anti-SARS coronavirus 3C-like protease effects of *Isatis indigotica* root and plant-derived phenolic compounds. *Antiviral research*, *68*(1), 36-42.

[46] Ryu, Y. B., et al. (2010). Biflavonoids from *Torreya nucifera* displaying SARS-CoV 3CL(pro) inhibition. *Bioorgan Med Chem*, *18*(22), 7940-7947.

[47] Zhuang, M., et al. (2009). Procyanidins and butanol extract of Cinnamomi Cortex inhibit SARS-CoV infection. *Antiviral research*, *82*(1), 73-81.

[48] Wang, S. Q., et al. (2007). Virtual screening for finding natural inhibitor against cathepsin-L for SARS therapy. *Amino acids*, *33*(1), 129-135.

[49] Wang, X. Y., Jia, W., Zhao, A. H. & Wang, X. R. (2006). Anti-influenza agents from plants and traditional Chinese medicine. *Phytother Res*, *20*(5), 335-341.

[50] Jia, W. & Gao, W. Y. (2003) Is traditional Chinese medicine useful in the treatment of SARS? *Phytother Res*, *17*(7), 840-841.

[51] Chen, Z. & Nakamura, T. (2004). Statistical evidence for the usefulness of Chinese medicine in the treatment of SARS. *Phytother Res*, *18*(7), 592-594.

[52] Yu, W. C., Hui, D. S. C. & Chan-Yeung, M. (2004). Antiviral agents and corticosteroids in the treatment of severe acute respiratory syndrome (SARS). *Thorax*, *59*(8), 643-645.

[53] Zhao, F. C., Guo, K. J. & Li, Z. R. (2013). Osteonecrosis of the femoral head in SARS patients: seven years later. *Eur J Orthop Surg Tr*, *23*(6), 671-677.

In: Avian Influenza and SARS: Epidemiology ... ISBN: 978-1-63463-793-0
Editor: Hailey Benson © 2015 Nova Science Publishers, Inc.

Chapter 4

Severe Acute Respiratory Syndrome: Lessons in Planning for Pandemic Diseases

Nada Mohsen[1] *and Nehad Nasef*[1,*]
[1]Departement of Pediatrics, Faculty of Medicine,
University of Mansoura, Mansoura, Egypt

Abstract

Severe acute respiratory syndrome coronavirus (SARS-CoV) is a novel virus that caused the first major pandemic of the twenty-first century. It was first reported in China in late 2002. The outbreak of severe acute respiratory syndrome in 2002–2003 exacted considerable human and economic burdens on countries involved. There has been no infection reported since April 2004. In affected countries, 8098 people have had probable SARS diagnosed, 774 of whom have died, with case-fatality rate of around 10%.

The diagnosis of SARS can`t be based on clinical features only but it needs both epidemiologic, and laboratory criteria that have been set by the World health organization and the Centers for Disease Control and Prevention. The clinical course of SARS appears to follow a triphasic pattern. The major clinical features on presentation include persistent fever, chills, myalgia, dry cough, sore throat, coryza, and extrapulmonary

* Tel.: +2 0100122929, E-mail: nehad_nasef@mans.edu.eg.

manifestation especially diarrhea. The radiographic appearances of SARS show predominant involvement of the lung periphery and the lower zone. The treatment of SARS can be with appropriate antibiotic coverage, antiviral agents and anti-inflammatory or immunomodulatory therapies. Mechanical ventilation can be used in cases with respiratory failure.

The prognostic factors associated with a poor outcome include age, chronic hepatitis B treated with lamivudine, high peak lactate dehydrogenase, high neutrophil count on presentation, presence of diabetes mellitus, or other comorbid conditions. A vaccine is not currently available in humans.

The prevention of SARS spread challenged government and health system due to physical stability of SARS-CoV in the environment, lack of protective immunity in the general population, and the absence of effective therapy or vaccination. Airborne precautions, inactivation of the virus using disinfectants, lung protective ventilation strategy, close observation to health care personnel caring for patient with suspected or confirmed SARS, and community spread controlling plans are our mean infection control weapons against SARS to prevent person-to-person transmission in future epidemics.

Abbreviations

Abbreviation	Whole word
ACE2	Angiotensin converting enzyme 2
CDC	Centers for Disease Control and Prevention
DC-SIGN	dendritic cell-specific intercellular adhesion molecule 3-grabbing non-integrin
E	Envelope protein
INF	Interferon
M	Membrane protein
MERS-COV	Middle East Respiratory Syndrome coronavirus
N	nucleocapsid protein
nsPs	non-structural proteins
RT-PCR	Reverse transcriptase- polymerase chain reaction
S	spike protein
SARS-COV	Severe acute respiratory syndrome coronavirus
siRNA	Small interfering RNAs
WHO	World Health Organization

Introduction

Severe acute respiratory syndrome (SARS) is a highly contagious respiratory disease caused by a SARS coronavirus (SARS-CoV) [1]. It is a novel virus that caused the first major pandemic of the twenty-first century [2]. Although the death rate is low in comparison with fatalities during previous pandemics; rapid spread, severity, human and economic burden, the potential misuse of the virus as a biological weapon, as well as unresolved origins and potential reservoirs are all contribute to continue research on different aspects of the SARS-CoV for fear of another SARS outbreak [3, 4].

Microbiology

SARS-COV is a large (about 29.700 nucleotides), single-stranded, positive-sense RNA enveloped virus that belongs to Coronaviruses which are members of Coronaviridae family. The genome is predicted to consist of at least 14 functional opening reading frames (ORFs) that encode 3 classes of proteins: two large polyproteins which are cleaved into 16 non-structural proteins (nsPs), 4 structural proteins, and 8 accessory proteins.

The first 2/3 of the SARS-COV genome consists of the viral replicase genes that encodes the 16 nsPs required mainly for viral RNA synthesis [5, 6]. The nsP12, nsP13, nsP5 and the nsP9 are certainly critical components for viral replication, so they can be the targets for the development of SARS-COV inhibitors [7, 8]. The last 1/3 part encodes for 4 structural proteins (spike (S), membrane (M), envelope (E), and nucleocapsid (N) which are essential for viral assembly) and 8 accessory proteins which their functions are poorly characterized [6, 8].

Molecolur Pathogenesis and Its Clinical Impacts

1. Functional Proteins

i. SARS Spike Protein Fusion Core

Infection of host cells by SARS-CoV is initiated by the interaction of the viral S-protein with receptors on the cell surface; this is an essential step for viral replication. The recognition of structural organization and membrane fusion activity of the S-protein will facilitate the generation of more potent inhibitors and vaccines candidates.

The SARS-CoV S protein can be subdivided into an N-terminal half (S1) and C-terminal half (S2) [9]. S1 is responsible for tissue tropism through binding to cellular receptors [10]. S2 is responsible for cell entry as it contains an internal fusion peptide allowing virus and host cell membrane fusion. Understanding the structure and function of fusion proteins allows the synthesis of several membrane fusion inhibitory peptides which block the viral entry [11].

ii. SARS Nucleocapsid Protein RNA-Binding Domain

The N protein can bind to the genomic RNA via a leader sequence leads to the formation of the helical ribonucleoprotein complex during viral assembly [12]. T-cells on the surface of infected cells can recognize antigenic peptides of the coronavirus N protein and the structure of these peptides could be used as a template for peptide-based vaccine design [13].

2. Cellular Receptors and Dendritic Cell

Because of the multiple implications for cell tropism, host range and inhibitor development, identification of the cellular receptors for SARS-CoV are the target of several laboratories researches.

i. ACE2 the Main Functional Cellular Receptor

Li et al. [10] and Wang et al. [14] reported that the metallopeptidase angiotensin converting enzyme 2 (ACE2) is a receptor for SARS-CoV. ACE2 dependent infection of organs other than the lung might contribute to SARS pathogenesis [15]. Identification of ACE2 on the surface of type 1 and type 2 alveolar epithelial cells, enterocytes of the small intestine and the brush border of the proximal tubular cells of the kidney using immunostaining techniques can explain the documented tissue tropism of SARS-CoV for some organs. However, lack of ACE2 protein expression in colonic enterocytes as well as the liver tissue, while SARS-CoV replication occurs there.

In contrast, whereas ACE2 is strongly expressed on the endothelial cells of arteries and veins of all tissues and the smooth muscle cells of the intestinal tract, there is no evidence of virus infection at any of these sites. These observations suggest that another cellular factor is required for successful virus infection [15].

ii. Dendritic Cells

Binding of SARS-CoV S-protein to dendritic cell-specific intercellular adhesion molecule 3-grabbing non-integrin (DC-SIGN) other than the receptors does not enable entry but can enhance viral infection and promote their spread within infected individuals in addition to internalization of SARS-CoV by denderitic cells might provide the virus with a means of immune escape [16].

iii. Human CD209L Receptors

It is thought that human CD209L receptors which are also called L-SIGN (for liver/ lymph node-specific) that are expressed in human lung in type II alveolar cells (the target cells for SARS-CoV infection) can mediate infection by SARS-CoV but less efficiently than ACE2 receptor [17].

3. Entry into Target Cells

After the attachment of virus to the host cells, the virus enters the cell either through pH-dependent receptor-mediated endocytosis or through direct membrane fusion.

i. Cell Entry by Direct Cell Fusion

The direct cell entry of SARS-COVs is mediated by S protein through S1 and S2 [18].

ii. Cell Entry by Endocytosis

The S protein not only mediates entry of SARS-CoV through direct membrane fusion but also allows entry through a pH-dependent endocytosis fashion. Cathepsin L is a pH-dependent cysteine protease which helps in cleavage of S protein or ACE2 and facilitates fusion of the viral membrane and the vesicle membrane with its maximal activity in an acidic milieu [19].

4. The Function of the Immune System

i. Innate Immune Response

Researches on innate immune responses to SARS-COVs detect that infection with SARS-COV results in decrease the production of type I interferons and increases the production of chemokines by infected organs which may be responsible for massive immune cell infiltrations of these organs and the dysregulation of adaptive immunity [20].

ii. The Adaptive Immune Response

T cells play important role in adaptive immunity against viral infections in vivo. In the case of acute phase of SARS-CoV infection, patients suffer from lymphopenia with a rapid decrease in both CD4 T which help in the production of virus-specific antibodies by B cells and CD8 T cells which can kill virus-infected host cells. Detection of neutralizing antibodies in SARS patients suggests that humoral immunity plays an important role in the elimination of the virus [21, 22].

5. The Role of Apoptosis

Fielding et al., [23]; Law et al., [24]; and Yuan et al., [25] did in-vitro experiments which indicated that over expression of certain nsPs may induce apoptosis in several cell types. Apoptosis can have two opposite roles on the pathogenicity of viral infection, clear off the virus in the infected cells and also can assist virus dissemination by the release of viral particles. Apoptosis in case of SARS-COVs have a deleterious pathogenic role, leading to severe tissue damage [26].

6. Does Genetic Affect the Pathogenesis of SARS?

Ip et al. [27] and Ng et al., [28] found that certain HLA haplotypes were associated with a higher susceptibility to SARS infection. L-SIGN homozygote individuals have a lower susceptibility to SARS infection [29].

Epidemiology

In November 2002, SARS emerged as an outbreak of atypical acute community acquired pneumonia. The primary case may have been a man in Foshan, China who was hospitalized and treated in an intensive care unit then he recovered but transmitted infection to his wife and other family members. Multiple cities around Guangzhou in Guangdong Province reported sporadic cases of an atypical pneumonia between November 2002 and February 2003 causing about 305 cases of SARS identified [30, 31].

The global SARS pandemic was rapidly and globally disseminated when a medical professor from a teaching hospital in Guangzhou came to Hong Kong on 21 February 2003 and he transmitted the infection to other residents. The secondary cases spread the disease to hospitals in Hong Kong, and to other countries including Vietnam, Singapore, and Canada. SARS-COVs affected a total of 8096 patients in over 30 countries and 774 (case fatality 10%) of them died. The SARS epidemic ended in Taiwan on 5 July 2003 [32]. Sixty-six percent of the cases occurred in China, 22% in Hong Kong, 4% in Taiwan and 3% in both Singapore and Canada (Figure 1) [33].

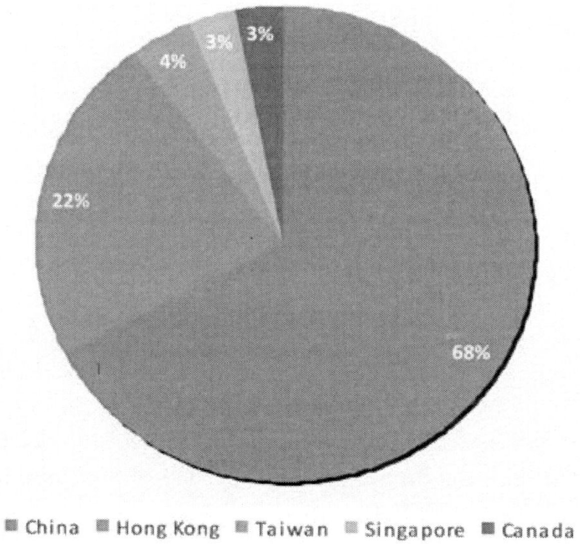

Figure 1. Incidence of SARS infected cases in the main five affected countries.

Dr. Carlo Urbani in Vietnam first reported a new global emerging virus respiratory threat to the WHO on February 28th 2003, resulting in a global alert notification on March 12th. Like many health care professionals who treated patients suffering from this new disease, Dr. Urbani became infected while treating patients and subsequently died from SARS infection [34].

Re-emergence of SARS had occurred from accidental laboratory exposures in Singapore, Taiwan, and Beijing and from recurrent animal-to-human transmissions in Guangzhou in late 2003 and early 2004 [35-38].

Hilgenfeld and Peiris tried to enumerate the chronology of events of SARS infection by time and place which were Partly based on WHO Western Pacific Region [39].

Transimission

The significant global, medical, economical, and social impact of SARS epidemic arouse the researchers to find the source of SARS-COVs. Phylogenetic analysis provided evidence for interspecies transmission events and interspecies exchange of genetic fragments raising the alarm that SARS-COV is a zoonotic virus that crossed the species barrier [40]. These studies identified the bats as the natural animal reservoir and civet and other mammals including raccoon dogs, and Chinese ferret badgers in wild life markets in South China as the intermediate amplifying hosts of SARS-CoV [41].

Phylogenetic analysis of the re-emergent viruses that affected four more patients with a SARS-like illness in late 2003 and early 2004 were more closely related to viruses found at that time in game-animal markets than to those that caused the SARS epidemic the year before [35]. Multiple factors may favor interspecies transmission as environmental, viral, and host factors.

Environmental factors had played a key role in the emergence of SARS-CoV. Overcrowding of the population as well as an increasing cultural demand for food and over-crowded game animals in unhygienic wild life markets of South China leaded to SARS-COVs cross the species barrier to humans [41].

Host cell receptor can determine whether inter species transmission of an emerging virus can occur or not, but many other cellular factors as the innate immune response, and metabolic characteristics of the host cell that may abort or facilitate the viral spread may have a role [42].

Contact with respiratory secretions was a significant risk factor for SARS transmission [43]. Exposure to body fluids of healthcare workers' eyes and mucous membranes was noticed to increase the risk of transmission [44]. Contaminated faeces proved to be important in at least one major community outbreak, Amoy Gardens in Hong Kong, in which over 300 patients were infected within the period of a few days [45].

The physical stability of SARS-CoV allowed transmission of SARS-CoV by indirect contact with contaminated environment. SARS-CoV can survive in the environment for 72–120 hours [46, 47] and the infectivity is remained for up to 6 days on a dried surface [48] while the respiratory specimens can be infectious for over 7 days at room temperature and SARS –COV can survive for 4 days in diarrheal stool samples with an alkaline pH [49]. Both the proximity and duration of contact with SARS patients may be associated with a higher risk of viral transmission [50, 51].

Clinical Features

It has been suggested that the clinical course of SARS passes into three phases. In the first phase, there is viral replication and patients suffer from the systemic symptoms. In the second phase, viral loads start to decrease but antibodies levels increase. However, pneumonia and immunopathological injury can also occur in this phase. In the third phase, fatal cases of SARS progress to severe pneumonia and acute respiratory distress syndrome (ARDS) or complete resolution occurs (figure 2) [52, 53]. Lessler et al., [54] made a systematic review on nine public health important respiratory viral infections that estimated a median incubation period for SARS-COVs to be 4 days and 5% of SARS-COVs cases will develop symptoms by 1·5 days while 95% by 10·6 days after infection. Different clinical presentations included fever, chills, cough, headache, myalgia, fatigue and malaise, sore throat, rhinorrhea and dizziness were seen. These symptoms may be milder in children, and severe in elderly patients [55]. In a community cohort diarrhea was occurred at presentation in low percentage which increased after a mean of 7.5 days after onset of symptoms [52]. Patients with SARS infection may suffer spontaneous pneumomediastium and acute respiratory distress syndrome which requires mechanical ventilation [52, 56]. The radiographic features of SARS were ground-glass opacities and focal consolidations predominantly involved the peripheral and subpleural regions of the lower zones [57]. Extrapulmonary manifestations may be in the form of diastolic cardiac dysfunction [58], liver and renal impairment [59, 60] neuromuscular disorder [61], and neurological manifestations [62] Lymphopenia, leucopenia, thrombocytopenia were commonly observed [63].

Diagnostic Criteria

The World Health Organization (WHO) and the Centers for Disease Control and Prevention (CDC) suggested diagnostic criteria for SARS during the initial phase of the epidemic (Tables 1, 2).

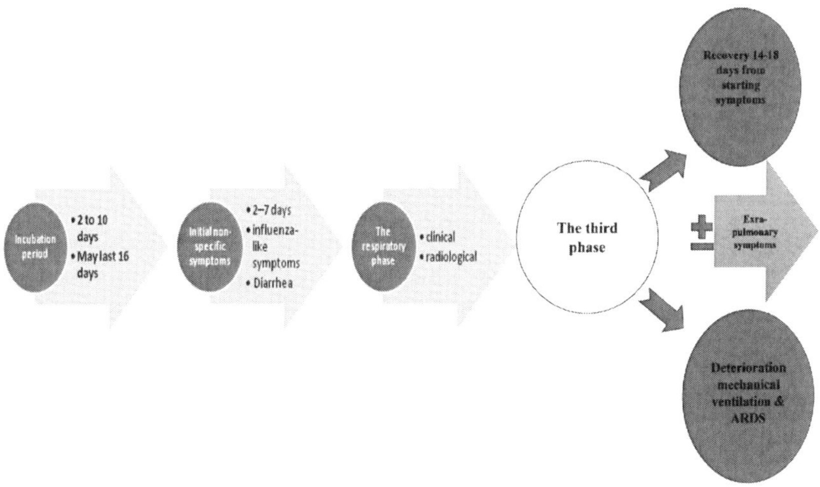

Figure 2. Clinical phases of SARS infection.

The overall accuracy of the WHO guidelines for identifying suspected SARS was found to be 83% with an negative predictive value of 86% [64].

Laboratory Diagnosis

A combination of classical methods could identify the SARS-COV as cell culture, electron microscopy, immunofluorescence and molecular biology technique using RT-PCR [68]. The detection of the infectious virus can be done by inoculating suitable cell cultures (e.g., Vero cells) with patient specimens (such as respiratory secretions, blood or stool) and propagating the virus in vitro [69]. Amplified of SARS-CoV RNA from clinical specimens of respiratory secretions, stool, blood, cerebrospinal fluid, and urine can be done by PCR [70-72].

The sensitivity of SARS-coronavirus detection improved to approximately 80% of nasopharyngeal aspirate specimens collected from day 1–3 of disease onset with 100% specificity by extraction of RNA samples from a larger volume of nasopharyngeal aspirate and by applying quantitative real-time RT-PCR technologies [73].

In addition, the sensitivity can be increased also by testing multiple serial samples collected from the same patient [72]. Faecal samples had very high viral loads at the end of the first week of illness, so it were the specimen of choice during the second week of disease [74].

Table 1. WHO diagnostic criteria for SARS

The WHO diagnostic criteria for SARS
I) Suspected case was defined as a person presenting after 1 November 2002 With*
A history of fever >38°C, with cough or difficulty breathing. And Close contact with a suspected or probable case of SARS, or had a history of traveling to or residing in an area with transmission of SARS within 10 days before the onset of symptoms. Or A person with an unexplained acute respiratory illness resulting in death, with the clinical criteria of suspected SARS and epidemiological exposure similar to that described above, but on whom no autopsy was performed.
II) A probable case of SARS was defined as a suspected case With*
Chest X-ray evidence of infiltrates consistent with pneumonia or acute respiratory distress syndrome. And A positive test result for SARS-CoV by one or more laboratory diagnostic assays. And/or Autopsy findings consistent with the pathology of ARDS, without an identifiable cause
III) A laboratory case definition for the diagnosis of a re-emergence of SARS#

Table 1. (Continued)

A person with clinically suggestive symptoms and signs with *one or more* positive laboratory findings including Reverse transcriptase- polymerase chain reaction(RT-PCR) positive for SARS-CoV, using a validated method From at least two different clinical specimens. Or The same clinical specimen collected on two or more occasions during the course of illness. Or Two different assays against different gene targets. Or Repeated RT-PCR using a new RNA extract from the original clinical sample on each occasion of testing. Seroconversion by enzyme-linked immunosorbent assay or indirect fluorescent antibody assay.
The WHO diagnostic criteria for SARS
Isolation in cell culture of SARS-CoV from any specimen and with RT-PCR confirmation using a validated method.
IV) Reclassification of cases*
As SARS is a diagnosis of exclusion. A patient should always be managed as clinically appropriate, regardless of their case status. - A case initially classified as suspect or probable: when an alternative diagnosis can fully explain the illness, SARS should be excluded after carefully considering the possibility of co-infection. - A suspect case who fulfils the probable case definition after investigation should be reclassified as (probable). - A suspect case with a normal CXR should be treated as appropriate and monitored for 7 days. If recovery is inadequate, it should be re-evaluated by CXR. - Those suspect case that recovery is adequate but whose illness cannot be fully explained by an alternative diagnosis should remain as (suspect). - A suspect case who dies, on whom no autopsy is done, should remain classified as (suspect). However, if this case is identified as being part of a chain transmission of SARS, the case should be reclassified as (probable). - If an autopsy is conducted and no pathological evidence of RDS is found, the case should be discarded.

* World Health Organization. Case definitions for surveillance of severe acute respiratory syndrome (SARS). 2003. Available at http://www.who.int/csr/sars/casedefinition/en/print.html [65].

WHO. Alert, verification and public health management of SARS in the post-outbreak period. Available on-line August. 2003;14 [66].

Various methods were developed for the detection of antibodies produced in response to infection with SARS-CoV. Serodiagnosis were based on the indirect immunofluorescence assay of virus-infected cells which yields a positive result from about day 10 after and an enzyme-linked immunosorbent assay based on viral cell extracts which yields positive results at around day 21 after the onset of illness [69, 75]. Detection of IgM or IgA subclass-specific antibody usually occurred in weeks 2 or 3 of illness, and a few patients seroconverted as late as 28 days after disease onset [52]. Visualisation of a virus-like agent in clinical material from SARS cases by electron microscopy provided poor diagnostic modality [76].

Treatment

The rapid spread, the serious illness, and the likehood of future occurrences of both SARS-CoV and related viruses highlight the need for rapid development of effective anti-viral drugs against emerging viral infections and necessitates continuous research for identification of antiviral therapies.

Table 2. The CDC case definition of SARS

The CDC case definition of SARS
A combination of clinical, epidemiological and laboratory criteria with the following exclusion criteria Cases in which an alternative diagnosis could fully explain the illness. The convalescent phase serum sample obtained >28 days after symptom onset was negative for antibody to SARS-CoV. The case was reported on the basis of contact with an index patient who was subsequently excluded as a case, providing other possible epidemiological exposure criteria were not available.

Centre of Disease Control and Prevention. Revised US surveillance case definition for severe acute respiratory syndrome (SARS) and update on SARS cases--United States and worldwide, December 2003. MMWR Morbidity and mortality weekly report. 2003;52(49):1202 [67].

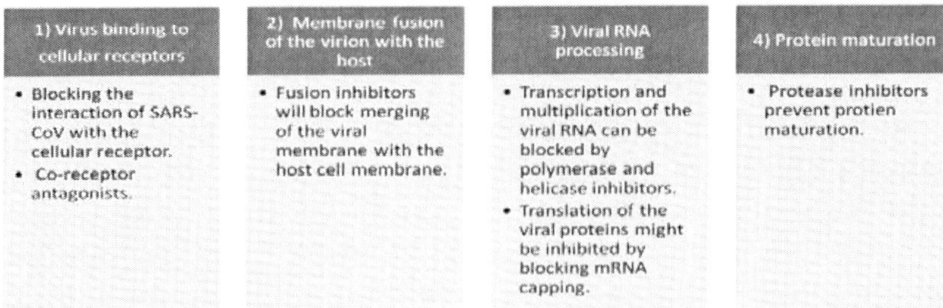

Figure 3. The SARS-CoV life cycle is vulnerable to therapeutic intervention in several places. Where (1) and (2) are entry inhibitors and (3) and (4) SARS-CoV enzymes as targets for antivirals. Kliger Y, Levanon EY, Gerber D. From genome to antivirals: SARS as a test tube. Drug discovery today. 2005;10(5):345-52 [77].

Drug development remains a challenge facing the molecular and structural biologists who suggested ways to interfere with the viral life cycle through virus identification, genome sequencing, and hopefully ends with an antiviral drug (figure 3). Researchers have served SARS as a test tube for novel approaches in therapy developed during the previous years battle against AIDS [77]. Ten years after the SARS outbreak, the world continues to be confronted by novel emerging disease threats leading the researchers to serve Middle East respiratory syndrome coronavirus again as a test tube for applications of subsequent understanding of the SARS-CoV theraputic trials [39].

1. Primary Measures and Supportive Treatments

Proper isolation and the strict infection control measures prevent effectively further transmissions.

Empiric antibiotics that cover the organisms associated with any community acquired pneumonia of unclear aetiology, including agents with activity against both typical and atypical respiratory pathogens. Oxygen supplementation may be needed and severe cases seem to do better if intensive care including intubation and mechanical ventilation are commenced early [78].

The neuraminidase inhibitor oseltamivir was prescribed to some patients in Toronto and Singapore before the identification of SARS-CoV as the etiological agent of SARS but the prognosis was not affected [79, 80].

2. Entry Inhibitors

a. Blocking the Interaction between SARS-CoV and Its Cellular Receptors

A soluble form of ACE2, an ACE2 inhibitor and an antibody recognizing SARS-CoV S1 efficiently neutralized SARS-CoV in vitro, supporting that the ACE2-binding site of the S- protien is an attractive target for drug development [10, 81, 82].

b. Fusion Inhibitors

The heptad region in the S-protein of SARS-CoV involves in the mechanism mediating fusion between the virus and host-cell membranes and its blocking inhibited viral replication in vitro [18].

Table 3. Studies of different regimen of ribivirin

Referance	Study Type	Regimen	Outcome
Morgenstern et al., 2005 [120]	Retrospective clinical trial	Ribivirin alone	No obvious clinical benefit
Avendano et al., 2003 [121]	Expiremental trial on pig's kidney cells	Ribivirin alone	Good activity
Avendano et al., 2003, Bitnun et al., 2003, Chiu et al., 2003, [76, 80, 122-124]	Clinical trials	Ribavirin as initial therapy.	Mixed results
Hsu et al., 2003 [80]	Clinical trial	Ribavirin versus Oseltamivir	No obvious response to late use of Ribavirin.
Chen et al., 2004 [125]	Experimental trial.	Ribavirin + interferon	Addition of interferon showed synergistic effect.

Table 3. (Continued)

Referance	Study Type	Regimen	Outcome
Chu et al., 2004 [87]	Experimental trial.	Ribavirin + Lopinavir + Steroid compared to Ribavirin + Steroid	Addition of Lopinavir showed synergistic effect
Que et al., 2003 [86]	Clinical trial.	Ribavirin + Lopinavir/ritonavir as initial or rescue therapy versu Ribavirin + steroid.	Lopinavir/ritonavir reduced the overall death and intubation rates.
Booth et al., 2003, Sung et al., 2004, Ooi et al., 2003 [52, 79, 84, 126]	Clinical trials	Ribavirin + Corticosteroid therapy with or without pulse steriod at clinical deterioration.	Mixed results

3. SARS-CoV Enzymes as Targets for Antivirals

a. RNA-Dependent RNA Polymerase

- The most commonly used agent was the broad spectrum antiviral ribavirin (table 4). However, the role of ribavirin remained uncertain [83]. Different regimens of ribavirin have been used [52, 79, 84]:
- Intravenous formulation of 8 mg/kg every 8 h for 14 days;
- Intravenous formulation of 8 mg/kg every 8 h for 5 days, followed by oral formulation of 1200 mg every 8 h, for a total of 10–14 days;
- Intravenous loading dose of 2 g, followed by intravenous formulation of 1 g every 6 h for 4 days, then 500 mg every 8 h for 3 days; or
- Oral formulation of 2.4 g for one dose, followed by 1.2 g every 8 h for 12 days.

Knowles et al., [85] studied the ribavirin side effects in a cohort of 110 patients in Toronto and dose-related hemolytic anemia was observed in 61%

of patients, whereas hypocalcaemia and hypomagnesaemia was reported in 58% and 46% respectively.

b. Protease Inhibitors

A combination of lopinavir and ritonavir had been used as either initial therapy or rescue therapy along with ribavirin during SARS epidemic and that combination showed synergistic effect with ribavirin [86, 87]. The nelfinavir, Calpain inhibitor III and Calpain inhibitor IV are other protease inhibitors which have in vitro activity against SARS-CoV and may be considered for the treatment of SARS [88].

c. Helicase Inhibitors

Sarafianos et al., [89] recognize that NTPase/helicase that is known as nspl3 are valid antiviral targets.

Helicases are motor proteins that unwind double stranded nucleic acids and may also be involved in protein-protein interactions and interactions with host factors.

Enzyme kinetic studies on the mechanism of nspl3 inhibition showed that SSYA10-001 specifically blocks nspl3 through a novel mechanism and that SSYA10-001 block viral replication but does not interfere with the functions of cellular enzymes that process nucleic acid or ATP.

Table 4. Studies of different regimen of corticosteroid

Referance	Study type	Regimen	Outcome
Chen et al., 2006 [127]	Retrospective clinical trial	The use of steroid in SARS patients	Corticosteroids lower the overall mortality and shorten hospitalization stay in the critically ill SARS patients.
Ho et al., 2003 [128]	Clinical trial	High-dose pulse methyl prednisolone (500 mg/d for 5 to 7 consecutive days, then tapered to 20mg/d over a total course of 21 days) versus Low-dose non-pulse corticosteroid (2 to 3 mg/kg/d of methylprednisolone)	Patients received high dose pulse corticosteroid had less oxygen requirement, better radiographic outcome, and less likelihood of requiring subsequent high-dose pulse corticosteroid therapy than their counter parts.

Table 4. (Continued)

Referance	Study type	Regimen	Outcome
Zhao et al., 2003 [129]	Clinical trial	Compared the clinical outcome of four treatment protocols Groups A: Ribavirin + Cefoperazone/Sulbactam Groups B: Fluoroquinolone + Azithromycin + IFN-α Group C: Quinolone + Azithromycin + recombinant IFN-α + methyl prednisolone (80–160 mg/ day for 2–3 days). Group D: Levofloxacin + Azithromycin + recombinant IFN-α + high-dose methyl prednisolone for 5–14 days (160–1000 mg/day)	Early high-dosage steroids resulted in best outcomes and no death as compared with low-dose steroid therapy.
Sung et al., 2004 [84]	Clinical trial	The methylprednisolone (500 mg/d) was administered if the patient deteriorated despite initial treatment	High-dose pulse methylprednisolone was associated with clinical improvement.

4. Immunomodulators

a. Corticosteriods

In addition to standard antiviral agents, some researchers reported that the steroids were associated with improvements of SARS patients based on the role of inflammatory response in the clinical progression of SARS. Evidence suggested that the clinical worsening during the mid-phase of SARS may be a result from the immunopathologic damage which can be reduced by steroids [90], in addition to elevated cytokine profile of SARS patients for at least 2 weeks after disease onset where corticosteroid can reduce significantly IL-8, monocyte chemoattractant protein-1, and IP-10 concentrations from days 5 to 8 after treatment [91].

Various regimens of corticosteroid therapy were used (table 5). The standard treatment protocol for adult SARS patients, comprising a tailing dose of intravenous methylprednisolone from 1 mg/kg every 8 h to oral prednisolone 0.25 mg/kg throughout a course of 21 days [78].

The use of steroid in critically ill SARS patients improved pulmonary oxygenation, reducing the need for intubation, case fatality, and pulmonary fibrosis. However, it is well known that corticosteroid therapy is a double-edged sword. Lee et al., [92] found that early corticosteroid treatment might be associated with a higher subsequent plasma viral load. Lv et al., [93] did a longitudinal follow up of 71 patients (mainly healthcare workers) who had been treated with corticosteroid of which 39% developed avascular necrosis of the hips within 3–4 months after starting treatment, and 58% of 71 patients had avascular necrosis after 3 years of follow up. Zhang and coworkers [94] correlated the number of osteonecrotic lesions with the dosage of corticosteroid where a peak dose of more than 200 mg or a cumulative methylprednisolone equivalent dose of more than 4000 mg were significant risk factors for multifocal osteonecrosis.

b. Convalescent Plasma

Using sera from people convalescing from SARS to treat SARS patients was proved to be effective, making the neutralizing antibody one of therapeutic strategy. Mair-Jenkins et al. [95] conducted a systematic review to evaluate the clinical effectiveness of convalescent plasma or serum for the treatment of severe acute respiratory infections of viral aetiology, to help clinical management of Middle East Respiratory Syndrome coronavirus (MERS-CoV) and included 8 observational studies for use of convalescent plasma in SARS patients.

Table 5. Studies of different regimen of interferon

Referance	Study type	Regimen	Outcome
Loutfy et al., 2003 [108]	Clinical trial	Corticosteroid versus Corticosteroid + interferon alfacon-1.	Corticosteroid + INF alfacon-1 improved oxygenation requirement and radiograph abnormalities.

Table 5. (Continued)

Referance	Study type	Regimen	Outcome
Cinatl et al., 2003, Hensley et al., 2004, Ströher et al., 2004 [117, 130, 131]	Experimental trials	Interferons	Inconsistent results for interferon-ß1a and interferon-a2b
Tan et al., 2004 [132]	Experimental trials	INF-ß, INF-αn1, INF-αn3, and leukocytic INF-α on SARS-CoV in culture.	Inhibition of cytopathic effects of SARS-COV in culture.
Sainz et al., 2004 [133]	Experimental trials	INF-ß +INF-γ	Synergistically inhibition of SARS-CoV plaque formation by 30-fold and replication by 3000-fold at 24 h.
Haagmans et al., 2004 [134]	Experimental trials	Prophylactic treatment of SARS-CoV infected macaques with pegylated INF-α	Reduced viral replication and excretion, and viral antigen expression by type 1 pneumocytes.
Morgenstern et al., 2005 [121]	Experimental trials	Ribavirin + IFN-β in two human cell lines.	Inhibition of SARS-CoV replication in reduced concentrations compared to either single treatment.
Kumaki et al., 2008 [135]	Experimental trials	IFN-alfacon 1	IFN-alfacon 1 was shown to be more active against SARS-CoV when tested in human lung epithelial Calu-3 cells.

Available evidence suggests that convalescent plasma is likely to reduce mortality [96-102], shorten the hospital stay and discharge by the day 22 was greater after treatment [99, 100], decrease in viral load reported after treatment [103], peak in SARS -COVs antibodies levels occurred within three to five days after taking convalescent plasma [96], and no adverse events or complications reported after treatment [96, 100]. However, these results based on predominately observational uncontrolled studies, low quality studies at medium to high risk of bias so further well developed studies are needed (51).

c. Pentaglobin and High-Dose Intravenous Gamma Globulin

It is an IgM-enriched immunoglobulin preparation. Ho et al., [104] carried out a retrospective cohort analysis on 12 severely ill SARS patients who continued to deteriorate despite corticosteroid and ribavirin therapy where addition of pentaglobin showed significant improvement in radiographic scores and oxygen requirement and 10 patients made an uneventful recovery without adverse events attributable to pentaglobin administration, compared with the use of high-dose intravenous gamma globulin in Lew et al. [105] study who used high-dose intravenous gamma globulin (0.4 g/kg/day for 3 consecutive days) in combination with steriod in critically ill patients with SARS, which may be associated with deep venous thrombosis and pulmonary embolism.

d. Interferons

Interferons (INF) have been considered as the first line of defense against viral infections [106]. INF type I is divided into INF-α produced by virus-infected leukocytes and INF-β produced by virus-infected fibroblasts, or virus-infected epithelial cells. INFs-α can block the replication of several viruses and also activate the immune system [107]. INFs inhibit viral infection by inducing both innate and adaptive immune responses (eg, by altering the intracellular environment to restrict viral replication, and inducing signaling events that activate immune cell populations and thereby elicit an antiviral immune response) [108].

INF-alfacon 1 is a non-naturally occurring, novel synthetic recombinant type 1 INF-alpha and is tested in human trials [109]. TypI INFs had been shown to inhibit SARS-CoV replication in vitro most effectively in cells that had been pre-treated before virus inoculation [110]. There was no standard regime for INF use and adverse events were not well documented [108]. Several studies carried out on the use of interferons in SARS patients that revealed different results (table 5).

5. Others

a. Cytokines Inhibitors

Cytokines are protein molecules that are released by cells when activated by antigens and are believed to be involved in cell-to-cell communications, acting as enhancing mediators for immune responses through interaction with specifc cell surface receptors on leukocytes. Siber carried out invention which provided a therapeutically effective amount of an inhibitor of a SARS-associated infammatory cytokine in a pharmaceutically acceptable carrier [111].

b. Small Interfering RNAs

Small interfering RNAs (siRNA) mediate the specific degradation of mRNAs with identical sequences which targeted the replicase 1A region of SARS-CoV and demonstrated a strong inhibitory effect on replication of different strains of SARS-CoV in vitro. However, their clinical benefit awaits evaluation [112]. Chang et al., [113] demonstrated that siRNAs are effective both prophylactically and therapeutically in vivo through inhibition of SARS-CoVinfection in Rhesus macaques.

c. Cyclosporin A

Based on efforts that had been done to learn from HIV, many virologists noticed some sequence similarities between the SARS-CoV N protein and the capsid protein of HIV. The latter binds to cyclophilin A, a peptidyl prolyl cis/trans-isomerase of the host cell, which is incorporated into the HIV particle [114]. Chen et al., [115] confirmed the interaction between cyclophilin A and the N protein. Pfefferle et al., [116] detected a specific interaction between cyclophilin A and the SARS-CoV Nsp1and found cyclosporin A, an inhibitor of cyclophilin A to be a ''pan-coronavirus inhibitor''.

d. Traditional Chinese Medicines

In mainland China, traditional Chinese medicines were used in addition to one or more of the therapies described above, but no systematic studies were carried out for their effects except the anti-SARS-CoV activity of Glycyrrhizin, a compound found in liquorice was tested in cell culture but at relatively high concentrations [117].

It is suggested that their use can modulate or restore the immune system and eliminate the toxin as a result of SARS, but without randomized control trial data [118].

Momattin et al., [119] summarized 19 studies carried out on the therapieutic effectiveness of ribavirin, interferon, Lopinavir and ritonavir, and convalescent plasma with exclusion of corticosteroid studies in SARS-COV patients.

They conclouded that the use of ribavirin may improve the outcome, addition of lopinavir/ritonavir to ribavirin appeared to have a better outcome, studies on the use of interferon and convalescent plasma failed to yield potent therapeutic effects.

The reported treatment effects in SARS patients should be interpreted with caution due to the heterogeneity of the reviewed studies and there are still no proven or licensed therapies for any coronavirus infection.

The frustrating conclusion after ten years of excellent basic research on SARS-CoV is that we are still left with the therapies that showed only uncertain effects in the treatment of patients in 2003 [39].

Prognosis of SARS-COVs Patients

Old age, rigor at onset, secondary lower respiratory infections, high peak lactate dehydrogenase, high neutrophil count on presentation, use of steroid, and use of invasive ventilation were related to higher fatality in critically ill cases with SARS.

The invasive ventilation may increased fatality by increasing rate of complications or from its use in serious condition per se [63]. Some studies showed that pregnancy, chronic hepatitis B treated with lamivudine, the presence of diabetes or other co-morbid conditions were associated with a poor outcome [52, 79, 136, 137]. The quantitative PCR assays might provide critical prognostic information for clinical management as high viral load was associated with high mortality rate, the occurrence of diarrhea, and the requirement for intensive care [138, 139].

Prevention

There are no clinically approved vaccines or antiviral drugs available for SARS infection; thus, the development of effective therapeutic and preventive strategies that can be readily applied to new emergent strains is a research priority [140].

1. General Preventive Measures

Understanding the risk of transmission was the target of many clinical studies aiming at proper infection control measures. Performing high-risk patient care procedures such as intubation, chest physiotherapy, suctioning, use of bi-level positive airway pressure, high-flow mechanical ventilation, and nebulizer therapy had been associated with nosocomial transmission of SARS [141-143]. Hospital design with augmented air changes may be protective against nosocomial transmission of SARS [144].

Basic infection control measures with good hand hygiene practice can reduce the risk of community transmission. Droplet and contact precautions as N95 respirators, gloves, gowns, shoe covers, powered air purified respirators and goggles should be available in emergency room, intensive care unit, and isolation ward [145, 146]. Fever screening stations, triage of fever patients, separating SARS patients from other patients, separation of entrances and passage ways between patients and healthcare workers, and increase hand-washing facilities all demonstrated a protective effect for healthcare workers [147, 148]. Construction of standard negative-pressure isolation rooms was expedited, and the emergency room was moved outside the hospital complex for patient triage [149].

2. Immunization

a. Passive Immunization

Human monoclonal antibodies that neutralize SARS-CoV were shown to be protective for passive prophylaxis and immunotherapy (therapeutic vaccines) in laboratory animals [150].

b. Active Immunization (Vaccination)

The severe morbidity and mortality associated with SARS makes it imperative to develop an effective and safe vaccine to prevent its re-emergence and epidemics [151]. Some challenges in the development of a SARS-CoV vaccine face the virologists:

i. Animal models of SARS-CoV infection do not typically mimic human disease which stand as an obstacle in application of vaccines tested on them [152].

ii. The effectiveness of candidate vaccines have to be tested on large human populations in areas where the virus is endemic, but how will vaccine manufacturers test their efficacy if SARS fails to return? [153].
iii. Coronaviruses may induce a short-lived immunity. Long-term immunity studies for SARS-CoV are carried out [153].

The humoral and T-cell-mediated responses to animal coronaviruses may either exacerbate disease or cause new health problems which makes the safety concern of the use of some coronavirus vaccines doubtful [154].

Different strategies were used to generate SARS-CoV-specific vaccines.

a. Inactivated Virus Vaccines

The virion structure can be maintained by treating the SARS-COVs by the use of chemicals as Formalin, β-propiolactone [155, 156] or ultraviolet irradiation [157].

So, the virus cannot infect, propagate or cause disease. The vaccines were used with or without adjuvant.

Neutralizing antibodies was induced when the vaccine was administered in many studies, but also adverse side effects like T helper 2-stimulation and cytokine upregulation was observed after boosting with the inactivated virus [156, 158, 159].

b. Live-Attenuated Virus Vaccines

The SARS-CoV mutants lacking the envelope protein or the exonuclease were able to protect mice in challenge experiments. Nevertheless, recombination events with wild-type coronaviruses might cause reversion to virulence [160, 161]. They can stimulate T cell and B cell dependent immune responses [162].

c. Subunit Vaccines

The S and the N proteins were used as target antigens for SARS-CoV vaccines where the antigenic components are generated in vitro [163-165]. Subunit vaccines were not only able to demonstrate neutralizing antibodies and reduced SARS-CoV loads in S protein vaccinated mice, but also could induce B cell- and T cell-dependent immune responses [163]. Subbarao et al., [152] found that this subunit vaccine induced higher neutralizing antibody and more complete protection than that achieved by inoculation of mice with live SARS-CoV.

d. DNA Vaccines

DNA vaccines expressing the SARS-CoV S protein were able to induce neutralizing antibodies, whereas DNA vaccines that encode the N protein induced cell-mediated immunity but were not protective after high titer challenge. Furthermore, induction of delayed-type hypersensitivity was observed [166]. The combination of the DNA vaccine and the whole chemically inactivated vaccine can be used to increase the magnitude of the immune response, and can change the balance of humoral to cellular immune response [167].

e. Vector Vaccines

The used vector systems which deliver host viral genome after genetically modification can efficiently induce innate and B cell as well as T cell mediated immune responses. Different vectors can be used for delivering SARS-CoV S or N genes as parainfluenza virus [168], Adenoviruses [169] and vaccinia virus [163, 170] were used as vector systems. Especially, the modified vaccinia virus Ankara has widely been used for the vaccine development against human diseases [171].

f. Vaccines Produced in Plants

The S1 domain of SARS-CoV S protein has been expressed in tomato and low-nicotine tobacco plants to develop a safe, effective, and an inexpensive vaccine candidate. Plant-derived antigen induced systemic and mucosal immune responses in mice, which showed significantly increased levels of SARS-CoV specific IgA after oral ingestion of tomato fruits expressing S1 protein. Sera of mice parentrally primed with tobacco derived S1 protein revealed the presence of SARS-CoV-specific IgG [172].

Planning for Future SARS Pandemics

We should be prepared for future SARS pandemics as we are not certain if SARS will reappears. SARS may possibly spread again from SARS-infected animals, unrecognized transmission in humans, or persistent infection in humans. Preparation for future pandemics should take into consideration early detection of infected cases and proper containment to avoid viral spread without taking unnecessary steps that may cause gratuitous concern. The CDC established a SARS preparedness task force that includes clinical;

surveillance; laboratory; special studies; information technology; communication and education; and response and preparedness for community teams [173]. The following major steps are crucial in preparedness for future SARS pandemic:

1. Infection control measures

   Transmission of SARS occurs mainly by droplet infection as a result of direct contact with infected patients particularly among health care workers [174]. Accordingly, infection control measures in health care facilities to minimize viral transmission among health care workers and visitors. Mask fitting, with periodic check for staff proficiency in their use, should be done for all health care workers before they start their jobs [175].

   Standard infection control measures for health care workers when dealing with infected patients include hand washing with antisepsis, use of personal protective equipment when handling body excretions or secretions, appropriate handling of patient care equipment and soiled linen, prevention of sharp injuries, environmental cleaning and spills-management, appropriate handling of waste [176].

   In preparation for outbreaks, patients should be well-informed to practice infection control measures such as frequent hand washing, wearing masks, and proper handling of facial tissues to help contain respiratory droplets and secretions.

2. Laboratory measures

   Collaboration between laboratories in term of sharing findings so that they can all learn from each other's work, exchanging reagents, sharing specimens and tissues to conduct additional testing all help in minimizing disease spread, emergence of new stains, and the development of better and reliable diagnostics. The introduction of rapid advanced diagnostic tests, the sophisticated molecular and sequencing techniques had facilitated the investigation of outbreaks and pseudo outbreaks caused by unusual pathogens as SARS-COV [177, 178]. Although laboratory testing is important to confirm infection, it does not replace accurate case definitions, regular updates as information evolves for identifying and managing case [52].

   The fear from the escape of the virus from laboratories that contain live SARS –COVs pushes WHO to provide strict biohazard guidelines for the handling of dangerous virus and its secure storage. In addition, the fear from reemergence from the animal sources raise the world arousal about proper protective handling of animals and the role of closing the animal markets to temper the severity and scope of future outbreaks [179, 180].

The emergence of the SARS-CoV highlighted the critical need for maintaining active basic science research on virus families even if these viruses are associated with limited or benign disease outcomes in humans as a principle determinant against new emerging diseases [179]. The clinical experience from SARS which is a closely related virus to current Novel MERS-CoV had been used for rapid development of effective therapeutics against MERS-CoV [119]. The first cases of severe human pneumonia caused by influenza A (H5N1) detected between late 2003, and early 2004 were probably identified partly because of enhanced alertness after the SARS outbreak, and led to the recognition and reporting of disease in poultry [181].

3. Containment measures

General containment measures gain considerable importance in controlling SARS spread as vaccines and effective anti-viral drugs against SARS do not exist. General containment measures are very effective in preventing the transmission of SARS in the community and they include early detection of suspected cases, daily temperature monitoring during outbreaks, rapid isolation, effective contact tracing, quarantine of exposed persons, surveillance of fever clusters, and a dedicated ambulance service to transport all possible cases to a SARS-designated hospital [182].

In health care facilities, SARS patients should be isolated from non-SARS patients with proper assignment of staff who can be designated to care for SARS patients to reduce the risk of nosocomial transmission to non-SARS patients and health care workers. Restricting visitor access to healthcare facilities and limiting new elective admissions to hospitals where SARS cases were isolated helped to limit the number of newly infected individuals [182].

Provide guidance to travellers from and to countries with SARS outbreaks including issuance of travel alerts and advisories, distribution of health alert notices, response to arriving ill travelers, notification and follow-up of potentially exposed passengers on public conveyances, and arrival and departure restrictions on travelers helped in prevention of global SARS spread.

4. Communication measures:

   Rapid and accurate communications are important to ensure a systematic and coordinated response to any infectious disease outbreak. This can be achieved by strengthening the linkages and communication among clinicians, emergency rooms, infection control practitioners, hospitals, pharmaceutical companies, and public health personnel.

   Building an electronic notification system that allows medical practitioners and laboratory personnel to submit on-line notifications of infectious diseases in a rapid and secure fashion is crucial in times of disease outbreaks to control the spread of infectious disease [183].
5. Continuing Research

   There are many questions about the SARS virus which remain unanswered. Future research should continue to build up a scientific base to ensure that public health community is adequately prepared to meet the challenges of SARS. Future researches should continue to explore tools needed for early diagnosis of suspected cases and track global transmission of the virus; interrupt transmission and treat or prevent disease; understand the natural history of SARS; develop more effective prevention and treatment strategies; and adequately inform healthcare workers, policy makers, and the public about appropriate responses to SARS outbreak.

References

[1] Ksiazek T. G., Erdman D., Goldsmith C. S., Zaki S. R., Peret T., Emery S., et al. A novel coronavirus associated with severe acute respiratory syndrome. *New England Journal of Medicine*, 2003; 348(20): 1953-1966.

[2] Peiris J., Lai S., Poon L., Guan Y., Yam L., Lim W., et al. Coronavirus as a possible cause of severe acute respiratory syndrome. *The Lancet*, 2003; 361(9366):1319-1325.

[3] Organization W. H. China's latest SARS outbreak has been contained, but biosafety concerns remain—update 7. *WHO Media Centre*, 2004.

[4] Wang H., Rao S., Jiang C. Molecular pathogenesis of severe acute respiratory syndrome. *Microbes and infection*, 2007; 9(1): 119-126.

[5] Thiel V., Ivanov K. A., Putics Á., Hertzig T., Schelle B., Bayer S., et al. Mechanisms and enzymes involved in SARS coronavirus genome expression. *Journal of General Virology*, 2003; 84(9): 2305-2315.

[6] Ziebuhr J. Molecular biology of severe acute respiratory syndrome coronavirus. *Current opinion in microbiology*, 2004; 7(4):412-419.

[7] Bartlam M., Yang H., Rao Z. Structural insights into SARS coronavirus proteins. *Current opinion in structural biology*, 2005; 15(6): 664-672.

[8] Tan B., Loh J., Seah S., Koh V., Lim E., Liaw C., et al. Strategies adopted and lessons learnt during the severe acute respiratory syndrome crisis in Singapore. *Reviews in Medical Virology*, 2005; 15(1): 57-70.

[9] Supekar V. M., Bruckmann C., Ingallinella P., Bianchi E., Pessi A., Carfí A. Structure of a proteolytically resistant core from the severe acute respiratory syndrome coronavirus S2 fusion protein. *Proceedings of the National Academy of Sciences*, 2004; 101(52): 17958-17963.

[10] Li W., Moore M. J., Vasilieva N., Sui J., Wong S. K., Berne M. A., et al. Angiotensin-converting enzyme 2 is a functional receptor for the SARS coronavirus. *Nature*, 2003; 426 (6965): 450-454.

[11] Bosch B. J., Martina B. E., van der Zee R., Lepault J., Haijema B. J., Versluis C., et al. Severe acute respiratory syndrome coronavirus (SARS-CoV) infection inhibition using spike protein heptad repeat-derived peptides. *Proceedings of the National Academy of Sciences of the United States of America*, 2004; 101(22): 8455-8460.

[12] Lai M., Cavanagh D. The molecular biology of coronaviruses. *Advances in virus research*, 1997; 48:1-100.

[13] Blicher T., Kastrup J. S., Buus S., Gajhede M. High-resolution structure of HLA-A* 1101 in complex with SARS nucleocapsid peptide. *Acta Crystallographica Section D: Biological Crystallography*, 2005; 61(8): 1031-1040.

[14] Wang P., Chen J., Zheng A., Nie Y., Shi X., Wang W., et al. Expression cloning of functional receptor used by SARS coronavirus. *Biochemical and biophysical research communications*, 2004; 315 (2): 439-444.

[15] Hamming I., Timens W., Bulthuis M., Lely A., Navis G., van Goor H. Tissue distribution of ACE2 protein, the functional receptor for SARS coronavirus. A first step in understanding SARS pathogenesis. *The Journal of pathology*, 2004; 203(2): 631-637.

[16] Hofmann H., Pöhlmann S. Cellular entry of the SARS coronavirus. *Trends in microbiology*, 2004; 12(10): 466-472.

[17] Jeffers S. A., Tusell S. M., Gillim-Ross L., Hemmila E. M., Achenbach J. E., Babcock G. J., et al. CD209L (L-SIGN) is a receptor for severe

acute respiratory syndrome coronavirus. *Proceedings of the National Academy of Sciences of the United States of America*, 2004; 101(44): 15748-15753.

[18] Liu S., Xiao G., Chen Y., He Y., Niu J., Escalante C. R., et al. Interaction between heptad repeat 1 and 2 regions in spike protein of SARS-associated coronavirus: implications for virus fusogenic mechanism and identification of fusion inhibitors. *The Lancet*, 2004; 363 (9413): 938-947.

[19] Yang Z.-Y., Huang Y., Ganesh L., Leung K., Kong W.-P., Schwartz O., et al. pH-dependent entry of severe acute respiratory syndrome coronavirus is mediated by the spike glycoprotein and enhanced by dendritic cell transfer through DC-SIGN. *Journal of virology*, 2004; 78 (11): 5642-5650.

[20] Thiel V., Weber F. Interferon and cytokine responses to SARS-coronavirus infection. *Cytokine and growth factor reviews*, 2008; 19(2): 121-132.

[21] Wong R. S., Wu A., To K., Lee N., Lam C. W., Wong C., et al. Haematological manifestations in patients with severe acute respiratory syndrome: retrospective analysis. *Bmj.*, 2003; 326(7403): 1358-1362.

[22] Sui J., Li W., Roberts A., Matthews L. J., Murakami A., Vogel L., et al. Evaluation of human monoclonal antibody 80R for immunoprophylaxis of severe acute respiratory syndrome by an animal study, epitope mapping, and analysis of spike variants. *Journal of virology*, 2005; 79(10): 5900-5906.

[23] Fielding B. C., Tan Y.-J., Shuo S., Tan T. H., Ooi E.-E., Lim S. G., et al. Characterization of a unique group-specific protein (U122) of the severe acute respiratory syndrome coronavirus. *Journal of virology*, 2004; 78(14): 7311-7318.

[24] Law P. T., Wong C.-H., Au T. C., Chuck C.-P., Kong S.-K., Chan P. K., et al. The 3a protein of severe acute respiratory syndrome-associated coronavirus induces apoptosis in Vero E6 cells. *Journal of general virology*, 2005; 86(7): 1921-1930.

[25] Yuan X., Shan Y., Zhao Z., Chen J., Cong Y. G0/G1 arrest and apoptosis induced by SARS-CoV 3b protein in transfected cells. *Virol. J.*, 2005; 2(66): 198-0024.

[26] Yan H., Xiao G., Zhang J., Hu Y., Yuan F., Cole D. K., et al. SARS coronavirus induces apoptosis in Vero E6 cells. *Journal of medical virology*, 2004; 73(3): 323-331.

[27] Ip W. E., Chan K. H., Law H. K., Tso G. H., Kong E. K., Wong W. H., et al. Mannose-binding lectin in severe acute respiratory syndrome coronavirus infection. *Journal of Infectious Diseases*, 2005; 91(10): 1697-1704.

[28] Ng M. H., Lau K.-M., Li L., Cheng S.-H., Chan W. Y., Hui P. K., et al. Association of human-leukocyte-antigen class I (B* 0703) and class II (DRB1* 0301) genotypes with susceptibility and resistance to the development of severe acute respiratory syndrome. *Journal of Infectious Diseases*, 2004; 190(3): 515-518.

[29] Chan V. S., Chan K. Y., Chen Y., Poon L. L., Cheung A. N., Zheng B., et al. Homozygous L-SIGN (CLEC4M) plays a protective role in SARS coronavirus infection. *Nature genetics*, 2005; 38(1): 38-46.

[30] Control CfD, Prevention. Update: outbreak of severe acute respiratory syndrome--worldwide, 2003. *MMWR Morbidity and mortality weekly report*, 2003; 52(13): 269.

[31] Zhao G.-p. SARS molecular epidemiology: a Chinese fairy tale of controlling an emerging zoonotic disease in the genomics era. *Philosophical Transactions of the Royal Society B: Biological Sciences*, 2007; 362(1482): 1063-1081.

[32] Cheng V. C., Lau S. K., Woo P. C., Yuen K. Y. Severe acute respiratory syndrome coronavirus as an agent of emerging and reemerging infection. *Clinical microbiology reviews*, 2007; 20(4): 660-694.

[33] Cherry J. The chronology of the 2002–2003 SARS mini pandemic. *Paediatric respiratory reviews*, 2004; 5(4): 262-269.

[34] Peiris J., Guan Y., Yuen K. Severe acute respiratory syndrome. *Nature medicine*, 2004; 10:S88-S97.

[35] Liang G., Chen Q., Xu J., Liu Y., Lim W., Peiris J., et al. Laboratory diagnosis of four recent sporadic cases of community-acquired SARS, Guangdong Province, China. *Emerg. Infect. Dis.*, 2004; 10 (10): 1774-1781.

[36] Lim P. L., Kurup A., Gopalakrishna G., Chan K. P., Wong C. W., Ng L. C., et al. Laboratory-acquired severe acute respiratory syndrome. *New England Journal of Medicine*, 2004; 350(17): 1740-1745.

[37] Normile D. Mounting lab accidents raise SARS fears. *Science*, 2004; 304(5671):659-661.

[38] Normile D. Second lab accident fuels fears about SARS. *Science*, 2004; 303(5654):26-26.

[39] Hilgenfeld R., Peiris M. From SARS to MERS: 10 years of research on highly pathogenic human coronaviruses. *Antiviral research*, 2013; 100 (1): 286-295.

[40] Graham R. L., Baric R. S. Recombination, reservoirs, and the modular spike: mechanisms of coronavirus cross-species transmission. *Journal of virology*, 2010; 84(7): 3134-3146.

[41] Lau S. K., Woo P. C., Li K. S., Huang Y., Tsoi H.-W., Wong B. H., et al. Severe acute respiratory syndrome coronavirus-like virus in Chinese horseshoe bats. *Proceedings of the National Academy of Sciences of the United States of America*, 2005; 102(39): 14040-14045.

[42] Chan J. F.-W., To K. K.-W., Tse H., Jin D.-Y., Yuen K.-Y. Interspecies transmission and emergence of novel viruses: lessons from bats and birds. *Trends in microbiology*, 2013; 21(10):544-555.

[43] Teleman M., Boudville I., Heng B., Zhu D., Leo Y. Factors associated with transmission of severe acute respiratory syndrome among health-care workers in Singapore. *Epidemiology and infection*, 2004;132(05): 797-803.

[44] Raboud J., Shigayeva A., McGeer A., Bontovics E., Chapman M., Gravel D., et al. Risk factors for SARS transmission from patients requiring intubation: a multicentre investigation in Toronto, Canada. *PloS one*, 2010; 5(5):e10717.

[45] Chim S., Tsui S., Chan K., Au T., Hung E., Tong Y., et al. Genomic characterisation of the severe acute respiratory syndrome coronavirus of Amoy Gardens outbreak in Hong Kong. *The Lancet*, 2003; 362(9398): 1807-1808.

[46] Chan K., Peiris J., Lam S., Poon L., Yuen K., Seto W. The effects of temperature and relative humidity on the viability of the SARS coronavirus. *Advances in virology*, 2011; 2011.

[47] Pi S., Han J., Bi S.-L., Dong X. P. Stability of SARS coronavirus in human specimens and environment and its sensitivity to heating and UV irradiation. *Biomedical and Environmental Sciences*, 2003; 16:246-255.

[48] Rabenau H., Cinatl J., Morgenstern B., Bauer G., Preiser W., Doerr H. Stability and inactivation of SARS coronavirus. *Medical microbiology and immunology*, 2005; 194(1-2):1-6.

[49] Lai M. Y., Cheng P. K., Lim W. W. Survival of severe acute respiratory syndrome coronavirus. *Clinical Infectious Diseases*, 2005; 41 (7):e67-e71.

[50] Scales D. C., Green K., Chan A. K., Poutanen S. M., Foster D., Nowak K., et al. Illness in intensive care staff after brief exposure to severe

acute respiratory syndrome. *Emerging infectious diseases*, 2003; 9(10): 1205.

[51] Fowler R. A., Guest C. B., Lapinsky S. E., Sibbald W. J., Louie M., Tang P., et al. Transmission of severe acute respiratory syndrome during intubation and mechanical ventilation. *American journal of respiratory and critical care medicine*, 2004; 169(11):1198-1202.

[52] Peiris J., Chu C., Cheng V., Chan K., Hung I., Poon L., et al. Clinical progression and viral load in a community outbreak of coronavirus-associated SARS pneumonia: a prospective study. *The Lancet*, 2003; 361(9371): 1767-1772.

[53] Nicholls J. M., Poon L. L., Lee K. C., Ng W. F., Lai S. T., Leung C. Y., et al. Lung pathology of fatal severe acute respiratory syndrome. *The Lancet*, 2003; 361(9371):1773-1778.

[54] Lessler J., Reich N. G., Brookmeyer R., Perl T. M., Nelson K. E., Cummings D. A. Incubation periods of acute respiratory viral infections: a systematic review. *The Lancet infectious diseases*, 2009; 9(5):291-300.

[55] Khuan Yew C., Lee C. E., Ling M. L., Heng D. M. K., Yap S. G. Outbreak of severe acute respiratory syndrome in a tertiary hospital in Singapore, linked to an index patient with atypical presentation: epidemiological study. *Bmj.*, 2004; 328(7433):195.

[56] Chu C., Leung Y., Hui J., Hung I., Chan V., Leung W., et al. Spontaneous pneumomediastinum in patients with severe acute respiratory syndrome. *European Respiratory Journal*, 2004; 23(6): 802-804.

[57] Grinblat L., Shulman H., Glickman A., Matukas L., Paul N. Severe Acute Respiratory Syndrome: Radiographic Review of 40 Probable Cases in Toronto, Canada 1. *Radiology*, 2003; 228(3):802-809.

[58] Li S. S.-l., Cheng C.-w., Fu C.-l., Chan Y.-h., Lee M.-p., Chan J. W.-m., et al. Left ventricular performance in patients with severe acute respiratory syndrome a 30-day echocardiographic follow-up study. *Circulation*, 2003; 108(15):1798-1803.

[59] Chau T. N., Lee K. C., Yao H., Tsang T. Y., Chow T. C., Yeung Y. C., et al. SARS-associated viral hepatitis caused by a novel coronavirus: Report of three cases. *Hepatology*, 2004; 39(2):302-310.

[60] Chu K. H., Tsang W. K., Tang C. S., Lam M. F., Lai F. M., To K. F., et al. Acute renal impairment in coronavirus-associated severe acute respiratory syndrome. *Kidney international*, 2005; 67(2):698-705.

[61] Tsai L.-K., Hsieh S.-T., Chao C.-C., Chen Y.-C., Lin Y.-H., Chang S.-C., et al. Neuromuscular disorders in severe acute respiratory syndrome. *Archives of neurology*, 2004; 61(11):1669-1673.

[62] Lau S. K., Woo P. C., Wong B. H., Tsoi H.-W., Woo G. K., Poon R. W., et al. Detection of severe acute respiratory syndrome (SARS) coronavirus nucleocapsid protein in SARS patients by enzyme-linked immunosorbent assay. *Journal of clinical microbiology*, 2004; 42 (7):2884-2889.

[63] Lee N., Hui D., Wu A., Chan P., Cameron P., Joynt G. M., et al. A major outbreak of severe acute respiratory syndrome in Hong Kong. *New England Journal of Medicine*, 2003; 348(20):1986-1994.

[64] Rainer T. H., Cameron P. A., Smit D., Ong K. L., Hung A. N. W., Nin D. C. P., et al. Evaluation of WHO criteria for identifying patients with severe acute respiratory syndrome out of hospital: prospective observational study. *Bmj.*, 2003; 326(7403):1354-1358.

[65] Organization WH. Case definitions for surveillance of severe acute respiratory syndrome (SARS). Published 2003.

[66] Organization WH. Alert, verification and public health management of SARS in the post-outbreak period. *Available on-line August*, 2003;14.

[67] Control CfD, Prevention. Revised US surveillance case definition for severe acute respiratory syndrome (SARS) and update on SARS cases--United States and worldwide, December 2003. *MMWR Morbidity and mortality weekly report*, 2003; 52(49):1202.

[68] Drosten C., Preiser W., Günther S., Schmitz H., Doerr H. W. Severe acute respiratory syndrome: identification of the etiological agent. *Trends in molecular medicine*, 2003; 9(8):325-327.

[69] Berger A., Drosten C., Doerr H., Stürmer M., Preiser W. Severe acute respiratory syndrome (SARS)—paradigm of an emerging viral infection. *Journal of clinical virology*, 2004; 29(1):13-22.

[70] Hung E., Chim S., Chan P., Tong Y., Ng E., Chiu R., et al. Detection of SARS coronavirus RNA in the cerebrospinal fluid of a patient with severe acute respiratory syndrome. 2003.

[71] Li L., Wo J., Shao J., Zhu H., Wu N., Li M., et al. SARS-coronavirus replicates in mononuclear cells of peripheral blood (PBMCs) from SARS patients. *Journal of Clinical Virology*, 2003; 28(3):239-244.

[72] Yam W., Chan K., Poon L., Guan Y., Yuen K., Seto W., et al. Evaluation of reverse transcription-PCR assays for rapid diagnosis of severe acute respiratory syndrome associated with a novel coronavirus. *Journal of clinical microbiology*, 2003; 41(10):4521-4524.

[73] Poon L. L., Wong B. W., Chan K. H., Leung C. S., Yuen K. Y., Guan Y., et al. A one step quantitative RT-PCR for detection of SARS coronavirus with an internal control for PCR inhibitors. *Journal of clinical virology*, 2004; 30(3):214-217.

[74] Poon L. L., Chan K. H., Wong O. K., Cheung T. K., Ng I., Zheng B., et al. Detection of SARS coronavirus in patients with severe acute respiratory syndrome by conventional and real-time quantitative reverse transcription-PCR assays. *Clinical chemistry*, 2004; 50(1):67-72.

[75] Stöhr K. A multicentre collaboration to investigate the cause of severe acute respiratory syndrome. *The Lancet*, 2003; 361(9370):1730-1733.

[76] Poutanen S. M., Low D. E., Henry B., Finkelstein S., Rose D., Green K., et al. Identification of severe acute respiratory syndrome in Canada. *New England Journal of Medicine*, 2003; 348(20):1995-2005.

[77] Kliger Y., Levanon E. Y., Gerber D. From genome to antivirals: SARS as a test tube. *Drug discovery today*, 2005; 10(5):345-352.

[78] So L. K., Lau A. C., Yam L. Y., Cheung T. M., Poon E., Yung R. W., et al. Development of a standard treatment protocol for severe acute respiratory syndrome. *The Lancet*, 2003; 361(9369):1615-1617.

[79] Booth C. M., Matukas L. M., Tomlinson G. A., Rachlis A. R., Rose D. B., Dwosh H. A., et al. Clinical features and short-term outcomes of 144 patients with SARS in the greater Toronto area. *Jama*, 2003; 289(21):2801-2809.

[80] Hsu L.-Y., Lee C.-C., Green J. A., Ang B., Paton N. I., Lee L., et al. Severe acute respiratory syndrome (SARS) in Singapore: clinical features of index patient and initial contacts. *Emerging infectious diseases*, 2003; 9(6):713.

[81] Sui J., Li W., Murakami A., Tamin A., Matthews L. J., Wong S. K., et al. Potent neutralization of severe acute respiratory syndrome (SARS) coronavirus by a human mAb to S1 protein that blocks receptor association. *Proceedings of the National Academy of Sciences of the United States of America*, 2004; 101(8):2536-2541.

[82] Huentelman M. J., Zubcevic J., Prada J. A. H., Xiao X., Dimitrov D. S., Raizada M. K., et al. Structure-based discovery of a novel angiotensin-converting enzyme 2 inhibitor. *Hypertension*, 2004; 44(6):903-906.

[83] Cheng V. C., Chan J. F., To K. K., Yuen K. Clinical management and infection control of SARS: lessons learned. *Antiviral research*, 2013; 100 (2):407-419.

[84] Sung J., Wu A., Joynt G., Yuen K., Lee N., Chan P., et al. Severe acute respiratory syndrome: report of treatment and outcome after a major outbreak. *Thorax*, 2004; 59(5):414-420.

[85] Knowles S. R., Phillips E. J., Dresser L., Matukas L. Common adverse events associated with the use of ribavirin for severe acute respiratory syndrome in Canada. *Clinical infectious diseases*, 2003; 37(8):1139-1142.

[86] Que T., Wong V., Yuen K. Treatment of severe acute respiratory syndrome with lopinavir/ritonavir: a multicentre retrospective matched cohort study. *Hong Kong medical journal*, 2003; 9(6):399-406.

[87] Chu C., Cheng V., Hung I., Wong M., Chan K., Chan K., et al. Role of lopinavir/ritonavir in the treatment of SARS: initial virological and clinical findings. *Thorax*, 2004; 59(3):252-256.

[88] Yamamoto N., Yang R., Yoshinaka Y., Amari S., Nakano T., Cinatl J., et al. HIV protease inhibitor nelfinavir inhibits replication of SARS-associated coronavirus. *Biochemical and biophysical research communications*, 2004; 318(3):719-725.

[89] Sarafianos S. G., Adedeji A. O., Singh K. Suppression of Sars Replication by Sars Helicase Inhibitors. In: Google Patents; 2013.

[90] Jiang Y., Xu J., Zhou C., Wu Z., Zhong S., Liu J., et al. Characterization of cytokine/chemokine profiles of severe acute respiratory syndrome. *American journal of respiratory and critical care medicine*, 2005; 171 (8):850-857.

[91] Wong C., Lam C., Wu A., Ip W., Lee N., Chan I., et al. Plasma inflammatory cytokines and chemokines in severe acute respiratory syndrome. *Clinical and Experimental Immunology*, 2004; 136 (1):95-103.

[92] Lee N., Allen Chan K., Hui D. S., Ng E. K., Wu A., Chiu R. W., et al. Effects of early corticosteroid treatment on plasma SARS-associated< i> Coronavirus</i> RNA concentrations in adult patients. *Journal of Clinical Virology*, 2004; 31(4):304-309.

[93] Lv H., De Vlas S. J., Liu W., Wang T. B., Cao Z. Y., Li C. P., et al. Avascular osteonecrosis after treatment of SARS: a 3-year longitudinal study. *Tropical Medicine and International Health*, 2009; 14(s1):79-84.

[94] Zhang N.-F., Li Z., Wei H.-Y., Liu Z.-H., Hernigou P. Steroid-induced osteonecrosis the number of lesions is related to the dosage. *Journal of Bone and Joint Surgery, British Volume*, 2008; 90(9):1239-1243.

[95] Mair-Jenkins J., Saavedra-Campos M., Baillie K., Cleary P., Khaw F.-M., Lim W. S., et al. The effectiveness of convalescent plasma and

hyperimmune immunoglobulin for the treatment of severe acute respiratory infections of viral aetiology: a systematic review and exploratory meta-analysis. *Journal of Infectious Diseases*, 2014:jiu396.

[96] Yeh K.-M., Chiueh T.-S., Siu L., Lin J.-C., Chan P. K., Peng M.-Y., et al. Experience of using convalescent plasma for severe acute respiratory syndrome among healthcare workers in a Taiwan hospital. *Journal of Antimicrobial Chemotherapy*, 2005; 56(5):919-922.

[97] Wong V., Dai D., Wu A., Sung J. Treatment of severe acute respiratory syndrome with convalescent plasma. *Hong Kong Medical Journal*, 2003; 9(3): 199-201.

[98] Burnouf T., Radosevich M. Treatment of severe acute respiratory syndrome with convalescent plasma. *Hong Kong Med. J.*, 2003; 9 (4): 309.

[99] Cheng Y., Wong R., Soo Y., Wong W., Lee C., Ng M., et al. Use of convalescent plasma therapy in SARS patients in Hong Kong. *European Journal of Clinical Microbiology and Infectious Diseases*, 2005; 24 (1):44-46.

[100] Soo Y., Cheng Y., Wong R., Hui D., Lee C., Tsang K., et al. Retrospective comparison of convalescent plasma with continuing high-dose methylprednisolone treatment in SARS patients. *Clinical microbiology and infection*, 2004; 10(7):676-678.

[101] Nie Q.-H., Luo X.-D., Hui W.-L. Advances in clinical diagnosis and treatment of severe acute respiratory syndrome. *World Journal of Gastroenterology*, 2003; 9(6):1139-1143.

[102] Zhou X., Zhao M., Wang F., Jiang T., Li Y., Nie W., et al. [Epidemiologic features, clinical diagnosis and therapy of first cluster of patients with severe acute respiratory syndrome in Beijing area]. *Zhonghua yi xue za zhi*, 2003; 83(12):1018-1022.

[103] Hung I. F., To K. K., Lee C.-K., Lee K.-L., Chan K., Yan W.-W., et al. Convalescent plasma treatment reduced mortality in patients with severe pandemic influenza A (H1N1) 2009 virus infection. *Clinical Infectious Diseases*, 2011; 52(4):447-456.

[104] Ho J., Wu A., Lam B., Ooi G., Khong P., Ho P., et al. Pentaglobin in steroid-resistant severe acute respiratory syndrome. *The International Journal of Tuberculosis and Lung Disease*, 2004; 8(10):1173-1179.

[105] Lew T. W., Kwek T.-K., Tai D., Earnest A., Loo S., Singh K., et al. Acute respiratory distress syndrome in critically ill patients with severe acute respiratory syndrome. *Jama*, 2003; 290(3):374-380.

[106] Tong M. J., Reddy K. R., Lee W. M., Pockros P. J., Hoefs J. C., Keeffe E. B., et al. Treatment of chronic hepatitis C with consensus interferon: a multicenter, randomized, controlled trial. *Hepatology*, 1997; 26(3):747-754.

[107] Cinatl J., Morgenstern B., Bauer G., Chandra P., Rabenau H., Doerr H. Treatment of SARS with human interferons. *The Lancet*, 2003; 362 (9380):293-294.

[108] Loutfy M. R., Blatt L. M., Siminovitch K. A., Ward S., Wolff B., Lho H., et al. Interferon alfacon-1 plus corticosteroids in severe acute respiratory syndrome: a preliminary study. *Jama*, 2003; 290(24):3222-3228.

[109] Heathcote E. J., Keeffe E. B., Lee S. S., Feinman S. V., Tong M. J., Reddy K., et al. Re-treatment of chronic hepatitis C with consensus interferon. *Hepatology*, 1998; 27(4):1136-1143.

[110] Cinatl Jr. J., Michaelis M., Scholz M., Doerr H. W. Role of interferons in the treatment of severe acute respiratory syndrome. *Expert opinion on biological therapy*, 2004; 4(6):827-836.

[111] Siber G. R. Methods for treatment of severe acute respiratory syndrome (SARS). In: Google Patents; 2011.

[112] He M.-L., Zheng B., Peng Y., Peiris J. S., Poon L. L., Yuen K. Y., et al. Inhibition of SARS-associated coronavirus infection and replication by RNA interference. *Jama*, 2003; 290 (20):2665-2666.

[113] Chang Z., Babiuk L. A., Hu J. Therapeutic and Prophylactic Potential of Small Interfering RNAs against Severe Acute Respiratory Syndrome. *BioDrugs*, 2007; 21(1):9-15.

[114] Gamble T. R., Vajdos F. F., Yoo S., Worthylake D. K., Houseweart M., Sundquist W. I., et al. Crystal structure of human cyclophilin A bound to the amino-terminal domain of HIV-1 capsid. *Cell*, 1996; 87(7):1285-1294.

[115] Chen Z., Mi L., Xu J., Yu J., Wang X., Jiang J., et al. Function of HAb18G/CD147 in invasion of host cells by severe acute respiratory syndrome coronavirus. *Journal of Infectious Diseases*, 2005; 191(5): 755-760.

[116] Pfefferle S., Schöpf J., Kögl M., Friedel C. C., Müller M. A., Carbajo-Lozoya J., et al. The SARS-coronavirus-host interactome: identification of cyclophilins as target for pan-coronavirus inhibitors. *PLoS pathogens*, 2011; 7(10):e1002331.

[117] Cinatl J., Morgenstern B., Bauer G., Chandra P., Rabenau H., Doerr H. Glycyrrhizin, an active component of liquorice roots, and replication of SARS-associated coronavirus. *The Lancet*, 2003; 361(9374):2045-2046.

[118] Liu X., Zhang M., He L., Li Y. Chinese herbs combined with Western medicine for severe acute respiratory syndrome (SARS). *Cochrane Database Syst. Rev.*, 2012; 10.

[119] Momattin H., Mohammed K., Zumla A., Memish Z. A., Al-Tawfiq J. A. Therapeutic Options for Middle East Respiratory Syndrome Coronavirus (MERS-CoV)–possible lessons from a systematic review of SARS-CoV therapy. *International Journal of Infectious Diseases*, 2013; 17(10): e792-e798.

[120] Leong H. N., Ang B., Earnest A., Teoh C., Xu W., Leo Y. S. Investigational use of ribavirin in the treatment of severe acute respiratory syndrome, Singapore, 2003. *Tropical Medicine and International Health*, 2004; 9(8):923-927.

[121] Morgenstern B., Michaelis M., Baer P. C., Doerr H. W., Cinatl Jr. J. Ribavirin and interferon-β synergistically inhibit SARS-associated coronavirus replication in animal and human cell lines. *Biochemical and biophysical research communications*, 2005; 326(4):905-908.

[122] Avendano M., Derkach P., Swan S. Clinical course and management of SARS in health care workers in Toronto: a case series. *Canadian Medical Association Journal*, 2003; 168(13):1649-1660.

[123] Bitnun A., Allen U., Heurter H., King S. M., Opavsky M. A., Ford-Jones E. L., et al. Children hospitalized with severe acute respiratory syndrome-related illness in Toronto. *Pediatrics*, 2003; 112(4):e261-e261.

[124] Chiu W.-k., Cheung P. C., Ng K., Ip P. L., Sugunan V., Luk D. C., et al. Severe acute respiratory syndrome in children: Experience in a regional hospital in Hong Kong*. *Pediatric Critical Care Medicine*, 2003; 4(3): 279-283.

[125] Chen F., Chan K., Jiang Y., Kao R., Lu H., Fan K., et al. In vitro susceptibility of 10 clinical isolates of SARS coronavirus to selected antiviral compounds. *Journal of Clinical Virology*, 2004; 31(1):69-75.

[126] Ooi C. G., Khong P. L., Ho J. C., Lam B., Wong W. M., Yiu W. C., et al. Severe Acute Respiratory Syndrome: Radiographic Evaluation and Clinical Outcome Measures 1. *Radiology*, 2003; 229(2):500-506.

[127] Chen R.-c., Tang X.-p., Tan S.-y., Liang B.-l., Wan Z.-y., Fang J.-q., et al. Treatment of Severe Acute Respiratory Syndrome With

GlucosteroidsThe Guangzhou Experience. *Chest Journal*, 2006; 129(6): 1441-1452.

[128] Ho J. C., Ooi G. C., Mok T. Y., Chan J. W., Hung I., Lam B., et al. High–dose pulse versus nonpulse corticosteroid regimens in severe acute respiratory syndrome. *American journal of respiratory and critical care medicine*, 2003; 168(12):1449-1456.

[129] Zhao Z., Zhang F., Xu M., Huang K., Zhong W., Cai W., et al. Description and clinical treatment of an early outbreak of severe acute respiratory syndrome (SARS) in Guangzhou, PR China. *Journal of medical microbiology*, 2003; 52(8):715-720.

[130] Hensley L. E., Fritz E. A., Jahrling P. B., Karp C., Huggins J. W., Geisbert T. W. Interferon-β 1a and SARS coronavirus replication. *Emerging infectious diseases*, 2004; 10(2):317.

[131] Ströher U., DiCaro A., Li Y., Strong J. E, Aoki F., Plummer F., et al. Severe acute respiratory syndrome-related coronavirus is inhibited by interferon-α. *Journal of Infectious Diseases*, 2004; 189(7):1164-1167.

[132] Tan E., Ooi E. E., Lin C.-Y., Tan H. C., Ling A. E., Lim B., et al. Inhibition of SARS coronavirus infection in vitro with clinically approved antiviral drugs. *Emerg. Infect. Dis.*, 2004; 10(4):581-586.

[133] Sainz Jr. B., Mossel E. C., Peters C., Garry R. F. Interferon-beta and interferon-gamma synergistically inhibit the replication of severe acute respiratory syndrome-associated coronavirus (SARS-CoV). *Virology*, 2004; 329(1):11-17.

[134] Haagmans B. L., Kuiken T., Martina B. E., Fouchier R. A., Rimmelzwaan G. F., van Amerongen G., et al. Pegylated interferon-α protects type 1 pneumocytes against SARS coronavirus infection in macaques. *Nature medicine*, 2004; 10(3):290-293.

[135] Kumaki Y., Day C. W., Wandersee M. K., Schow B. P., Madsen J. S., Grant D., et al. Interferon alfacon 1 inhibits SARS-CoV infection in human bronchial epithelial Calu-3 cells. *Biochemical and biophysical research communications*, 2008; 371(1):110-113.

[136] Donnelly C. A., Ghani A. C., Leung G. M., Hedley A. J., Fraser C., Riley S., et al. Epidemiological determinants of spread of causal agent of severe acute respiratory syndrome in Hong Kong. *The Lancet*, 2003; 361 (9371):1761-1766.

[137] Wong S., Chow K., Swiet M. Severe acute respiratory syndrome and pregnancy. *BJOG: An International Journal of Obstetrics and Gynaecology*, 2003; 110(7):641-642.

[138] Cheng V., Hung I., Tang B., Chu C., Wong M., Chan K., et al. Viral replication in the nasopharynx is associated with diarrhea in patients with severe acute respiratory syndrome. *Clinical infectious diseases*, 2004; 38(4):467-475.

[139] Ng E. K., Ng P.-C., Hon K. E., Cheng W. F., Hung E. C., Chan K. A., et al. Serial analysis of the plasma concentration of SARS coronavirus RNA in pediatric patients with severe acute respiratory syndrome. *Clinical chemistry*, 2003; 49(12):2085-2088.

[140] Graham R. L., Donaldson E. F., Baric R. S. A decade after SARS: strategies for controlling emerging coronaviruses. *Nature Reviews Microbiology*, 2013; 11(12):836-848.

[141] Chen W.-Q., Ling W.-H., Lu C.-Y., Hao Y.-T., Lin Z.-N., Ling L., et al. Which preventive measures might protect health care workers from SARS? *BMC public health*, 2009; 9(1):81.

[142] Ofner-Agostini M., Gravel D., L. Clifford McDonald M., Lem M., Sarwal S., McGeer A., et al. Cluster of cases of severe acute respiratory syndrome among Toronto healthcare workers after implementation of infection control precautions: a case series. *infection control and hospital epidemiology*, 2006; 27(5):473-478.

[143] Christian M. D., Loutfy M., McDonald L. C., Martinez K. F., Ofner M., Wong T., et al. Possible SARS coronavirus transmission during cardiopulmonary resuscitation. *Emerging infectious diseases*, 2004; 10 (2): 287-293.

[144] Leitmeyer K. C., Anh B. H., Reynolds M. G., Montgomery J. M., Comer J. A., Horby P. W., et al. Lack of SARS transmission among public hospital workers, Vietnam. *Cough*, 2004; 10:16.11.

[145] Gopalakrishna G., Choo P., Leo Y. S., Tay B. K., Lim Y. T., Khan A. S., et al. SARS transmission and hospital containment. *Emerging infectious diseases*, 2004; 10(3):395.

[146] Dwosh H. A., Hong H. H., Austgarden D., Herman S., Schabas R. Identification and containment of an outbreak of SARS in a community hospital. *Canadian Medical Association Journal*, 2003; 168(11):1415-1420.

[147] Yen M.-Y., Lin Y.-E., Lee C.-H., Ho M.-S., Huang F.-Y., Chang S.-C., et al. Taiwan's traffic control bundle and the elimination of nosocomial severe acute respiratory syndrome among healthcare workers. *Journal of Hospital Infection*, 2011; 77(4):332-337.

[148] Yen M.-Y., Lin Y., Su I.-J., Huang F.-Y., Huang F.-Y., Ho M.-S., et al. Using an integrated infection control strategy during outbreak control to

minimize nosocomial infection of severe acute respiratory syndrome among healthcare workers. *Journal of Hospital Infection*, 2006;62(2): 195-199.

[149] Liu J. W., Lu S. N., Chen S. S., Yang K. D., Lin M. C., Wu C. C., et al. Epidemiologic study and containment of a nosocomial outbreak of severe acute respiratory syndrome in a medical center in Kaohsiung, Taiwan. *Infection control and hospital epidemiology*, 2006; 27(5):466-472.

[150] Gillim-Ross L., Subbarao K. Emerging respiratory viruses: challenges and vaccine strategies. *Clinical microbiology reviews*, 2006; 19(4):614-636.

[151] Zhi Y., Wilson J. M., Shen H. SARS vaccine: progress and challenge. *Cell Mol. Immunol.*, 2005; 2(2):101-105.

[152] Subbarao K., McAuliffe J., Vogel L., Fahle G., Fischer S., Tatti K., et al. Prior infection and passive transfer of neutralizing antibody prevent replication of severe acute respiratory syndrome coronavirus in the respiratory tract of mice. *Journal of virology*, 2004; 78(7):3572-3577.

[153] Taylor D. R. Obstacles and advances in SARS vaccine development. *Vaccine*, 2006; 24(7):863-871.

[154] Zhong X., Yang H., Guo Z.-F., Sin W.-Y. F., Chen W., Xu J., et al. B-cell responses in patients who have recovered from severe acute respiratory syndrome target a dominant site in the S2 domain of the surface spike glycoprotein. *Journal of virology*, 2005; 79(6):3401-3408.

[155] Zhang C.-h., Lu J.-h., Wang Y.-f., Zheng H.-y., Xiong S., Zhang M.-y., et al. Immune responses in Balb/c mice induced by a candidate SARS-CoV inactivated vaccine prepared from F69 strain. *Vaccine*, 2005; 23 (24): 3196-3201.

[156] See R. H., Zakhartchouk A. N., Petric M., Lawrence D. J., Mok C. P., Hogan R. J., et al. Comparative evaluation of two severe acute respiratory syndrome (SARS) vaccine candidates in mice challenged with SARS coronavirus. *Journal of general virology*, 2006;87(3):641-650.

[157] Darnell M. E., Subbarao K., Feinstone S. M., Taylor D. R. Inactivation of the coronavirus that induces severe acute respiratory syndrome, SARS-CoV. *Journal of virological methods*, 2004; 121(1):85-91.

[158] Bolles M., Deming D., Long K., Agnihothram S., Whitmore A., Ferris M., et al. A double-inactivated severe acute respiratory syndrome coronavirus vaccine provides incomplete protection in mice and induces

increased eosinophilic proinflammatory pulmonary response upon challenge. *Journal of virology*, 2011; 85(23):12201-12215.

[159] Roper R. L., Rehm K. E. SARS vaccines: where are we? *Expert review of vaccines*, 2009; 8(7):887-898.

[160] Eckerle L. D., Becker M. M., Halpin R. A., Li K., Venter E., Lu X., et al. Infidelity of SARS-CoV Nsp14-exonuclease mutant virus replication is revealed by complete genome sequencing. *PLoS pathogens*, 2010; 6 (5):e1000896.

[161] Fett C., DeDiego M. L., Regla-Nava J. A., Enjuanes L., Perlman S. Complete protection against severe acute respiratory syndrome coronavirus-mediated lethal respiratory disease in aged mice by immunization with a mouse-adapted virus lacking E protein. *Journal of virology*, 2013; 87(12):6551-6559.

[162] Vignuzzi M., Wendt E., Andino R. Engineering attenuated virus vaccines by controlling replication fidelity. *Nature medicine*, 2008; 14(2): 154-161.

[163] Bisht H., Roberts A., Vogel L., Subbarao K., Moss B. Neutralizing antibody and protective immunity to SARS coronavirus infection of mice induced by a soluble recombinant polypeptide containing an N-terminal segment of the spike glycoprotein. *Virology*, 2005; 334(2):160-165.

[164] Gupta V., Tabiin T. M., Sun K., Chandrasekaran A., Anwar A., Yang K., et al. SARS coronavirus nucleocapsid immunodominant T-cell epitope cluster is common to both exogenous recombinant and endogenous DNA-encoded immunogens. *Virology*, 2006; 347(1):127-139.

[165] Liu S.-J., Leng C.-H., Lien S.-p., Chi H.-Y., Huang C.-Y., Lin C.-L., et al. Immunological characterizations of the nucleocapsid protein based SARS vaccine candidates. *Vaccine*, 2006; 24(16):3100-3108.

[166] Zhao P., Cao J., Zhao L.-J., Qin Z.-L., Ke J.-S., Pan W., et al. Immune responses against SARS-coronavirus nucleocapsid protein induced by DNA vaccine. *Virology*, 2005; 331(1):128-135.

[167] Zakhartchouk A. N., Liu Q., Petric M., Babiuk L. A. Augmentation of immune responses to SARS coronavirus by a combination of DNA and whole killed virus vaccines. *Vaccine*, 2005; 23(35):4385-4391.

[168] Bukreyev A., Lamirande E. W., Buchholz U. J., Vogel L. N., Elkins W. R., Claire M. S., et al. Mucosal immunisation of African green monkeys (< i> Cercopithecus aethiops)</i> with an attenuated parainfluenza virus

expressing the SARS coronavirus spike protein for the prevention of SARS. *The Lancet*, 2004; 363(9427):2122-2127.
[169] Gao W., Tamin A., Soloff A., D'Aiuto L., Nwanegbo E., Robbins P. D., et al. Effects of a SARS-associated coronavirus vaccine in monkeys. *The Lancet*, 2003; 362(9399):1895-1896.
[170] Chen Z., Zhang L., Qin C., Ba L., Christopher E. Y., Zhang F., et al. Recombinant modified vaccinia virus Ankara expressing the spike glycoprotein of severe acute respiratory syndrome coronavirus induces protective neutralizing antibodies primarily targeting the receptor binding region. *Journal of virology*, 2005; 79(5):2678-2688.
[171] Volz A., Sutter G. Protective efficacy of modified vaccinia virus Ankara in preclinical studies. *Vaccine*, 2013; 31(39):4235-4240.
[172] Pogrebnyak N., Golovkin M., Andrianov V., Spitsin S., Smirnov Y., Egolf R., et al. Severe acute respiratory syndrome (SARS) S protein production in plants: development of recombinant vaccine. *Proceedings of the National Academy of Sciences of the United States of America*, 2005; 102(25):9062-9067.
[173] Hughes J. M. The SARS response--building and assessing an evidence-based approach to future global microbial threats. *JAMA*, 2003; 290 (24):3251-3253.
[174] Tam D. K., Lee S., Lee S. S. Impact of SARS on avian influenza preparedness in healthcare workers. *Infection*, 2007; 35(5):320-325.
[175] Gomersall C. D., Tai D. Y., Loo S., Derrick J. L., Goh M. S., Buckley T. A., et al. Expanding ICU facilities in an epidemic: recommendations based on experience from the SARS epidemic in Hong Kong and Singapore. *Intensive Care Med.*, 2006; 32(7):1004-1013.
[176] World Health Organization. Avian influenza iiA, (H5N1) ihWiicgfh, http://www.who.int/ cfUFAa, csr/disease/avian_influenza/guidelines/ infectioncontrol1/en/index.html., 2006 AM.
[177] Cheng V. C.-c., Wong S. S.-y., Chen J. H.-k., Chan J. F.-w., To K. K.-w., Poon R. W.-s., et al. An unprecedented outbreak investigation for nosocomial and community-acquired legionellosis in Hong Kong. *Chin. Med. J.*, 2012; 125(23):4283-4290.
[178] To K. K., Fung A. M., Teng J. L., Curreem S. O., Lee K.-C., Yuen K.-Y., et al. Characterization of a Tsukamurella pseudo-outbreak by phenotypic tests, 16S rRNA sequencing, pulsed-field gel electrophoresis, and metabolic footprinting. *Journal of clinical microbiology*, 2013; 51(1): 334-338.

[179] Baric R. S. SARS-CoV: lessons for global health. *Virus research*, 2008; 133(1):1.

[180] Poon L., Guan Y., Nicholls J., Yuen K., Peiris J. The aetiology, origins, and diagnosis of severe acute respiratory syndrome. *The Lancet infectious diseases*, 2004; 4(11):663-671.

[181] Heymann D. L., Mackenzie J. S., Peiris M. SARS legacy: outbreak reporting is expected and respected. *The Lancet*, 2013; 381 (9869):779-781.

[182] Srinivasan A., McDonald L. C., Jernigan D., Helfand R., Ginsheimer K., Jernigan J., et al. Foundations of the severe acute respiratory syndrome preparedness and response plan for healthcare facilities. *Infect. Control Hosp. Epidemiol.*, 2004; 25(12):1020-1025.

[183] Lombardo J., Burkom H., Elbert E., Magruder S., Lewis S. H., Loschen W, et al. A systems overview of the Electronic Surveillance System for the Early Notification of Community-Based Epidemics (ESSENCE II). *J. Urban Health*, 2003; 80 (2 Suppl. 1):i32-42.

Index

A

B

C

D

E

F

G

H

I

N

O

P

T